Algebraic Groups and Differential Galois Theory

Algebraic Groups and Differential Galois Theory

Teresa Crespo
Zbigniew Hajto

Graduate Studies in Mathematics
Volume 122

American Mathematical Society
Providence, Rhode Island

2010 *Mathematics Subject Classification.* Primary 12H05, 13B05, 14A10, 17B45, 20G15, 34M35, 68W30.

Several sections and chapters are from the authors' book *Introduction to Differential Galois Theory*, Cracow University of Technology Press, Cracow, 2007, ISBN 978-83-7242-453-2, 94 pages.

For additional information and updates on this book, visit
www.ams.org/bookpages/gsm-122

Library of Congress Cataloging-in-Publication Data

Crespo, Teresa, 1955–
Algebraic groups and differential Galois theory / Teresa Crespo, Zbigniew Hajto.
p. cm. — (Graduate studies in mathematics ; v. 122)
Includes bibliographical references and index.
ISBN 978-0-8218-5318-4 (alk. paper)
1. Differential algebraic groups. 2. Galois theory. 3. Morphisms (Mathematics). I. Hajto, Zbigniew. II. Title.

QA247.4.C74 2011
515′.354—dc22

2010044378

∞ The paper used in this book is acid-free and falls within the guidelines established to ensure permanence and durability.
Visit the AMS home page at http://www.ams.org/

10 9 8 7 6 5 4 3 2 1 16 15 14 13 12 11

To the memory of Jerald Joseph Kovacic (1941–2009).

Contents

Preface

The aim of this book is to present the Galois theory of homogeneous linear differential equations. This theory goes back to the work of Picard and Vessiot at the end of the 19th century and bears their names. It parallels the Galois theory of algebraic equations. The notions of splitting field, Galois group, and solvability by radicals have their counterparts in the notions of Picard-Vessiot extension, differential Galois group, and solvability by quadratures. The differential Galois group of a homogeneous linear differential equation has a structure of linear algebraic group; hence it is endowed, in particular, with the Zariski topology. The fundamental theorem of Picard-Vessiot theory establishes a bijective correspondence between intermediate differential fields of a Picard-Vessiot extension and closed subgroups of its differential Galois group. Solvability by quadratures is characterized by means of the differential Galois group. Picard-Vessiot theory was clarified and generalized in the work of Kolchin in the mid-20th century. Kolchin used the differential algebra developed by Ritt and also built the foundations of the theory of linear algebraic groups. Kaplansky's book "Introduction to Differential Algebra" made the theory more accessible, although it omits an important point, namely the construction of the Picard-Vessiot extension. The more recent books by Magid and van der Put and Singer assume that the reader is familiar with algebraic varieties and linear algebraic groups, although the latter book compiles the most important topics in an appendix. We point out that not all results on algebraic varieties and algebraic groups needed to develop differential Galois theory appear in the standard books on these topics. For our book we have decided to develop the theory of algebraic varieties and linear algebraic groups in the same way that books on classical Galois theory include some chapters on group, ring, and field theories. Our text includes complete proofs, both of the results on algebraic geometry and algebraic groups which are needed in Picard-Vessiot theory and of the results on Picard-Vessiot theory itself.

We have given several courses on Differential Galois Theory in Barcelona and Kraków. As a result, we published our previous book "Introduction to Differential Galois Theory" [**C-H1**]. Although published by a university publishing house, it has made some impact and has been useful to graduate students as well as to theoretical physicists working on dynamical systems. Our present book is also aimed at graduate students in mathematics or physics and at researchers in these fields looking for an introduction to the subject. We think it is suitable for a graduate course of one or two semesters, depending on students' backgrounds in algebraic geometry and algebraic groups. Interested students can work out the exercises, some of which give an insight into topics beyond the ones treated in this book. The prerequisites for this book are undergraduate courses in commutative algebra and complex analysis.

We would like to thank our colleagues José María Giral, Andrzej Nowicki, and Henryk Żołądek who carefully read parts of this book and made valuable comments, as well as Jakub Byszewski and Sławomir Cynk for interesting discussions on its content. We are also grateful to the anonymous referees for their corrections and suggestions which led to improvements in the text. Our thanks also go to Dr. Ina Mette for persuading us to expand our previous book to create the present one and for her interest in this project.

Finally our book owes much to Jerry Kovacic. We will always be thankful to him for many interesting discussions and will remember him as a brilliant mathematician and an open and friendly person.

Both authors acknowledge support by Spanish Grants MTM2006-04895 and MTM2009-07024, Polish Grant N20103831/3261 and European Network MRTN-CT-2006-035495.

Barcelona and Kraków, October 2010

Teresa Crespo and Zbigniew Hajto

Introduction

This book has been conceived as a self-contained introduction to differential Galois theory. The self-teaching reader or the teacher wanting to give a course on this subject will find complete proofs of all results included. We have chosen to make a classical presentation of the theory. We refer to the Picard-Vessiot extension as a field rather than introducing the notion of Picard-Vessiot ring so as to keep the analogy with the splitting field in the polynomial Galois theory. We also refer to differential equations rather than differential systems, although the differential systems setting is given in the exercises.

The chapters on algebraic geometry and algebraic groups include all questions which are necessary to develop differential Galois theory. The differential Galois group of a linear differential equation is a linear algebraic group, hence affine. However, the construction of the quotient of an algebraic group by a subgroup needs the notion of abstract affine variety. Once we introduce the notion of geometric space, the concept of algebraic variety comes naturally. We also consider it interesting to include the notion of projective variety, which is a model for algebraic varieties, and present a classical example of an algebraic group which is not affine, namely the elliptic curve.

The chapter on Lie algebras aims to prove the equivalence between the solvability of a connected linear algebraic group and the solvability of its Lie algebra. This fact is used in particular to determine the algebraic subgroups of $\mathrm{SL}(2,\mathbb{C})$. We present the characterization of differential equations solvable by quadratures. In the last chapter we consider differential equations defined over the field of rational functions over the complex field and present classical notions such as the monodromy group, Fuchsian equations and hypergeometric equations. The last section is devoted to Kovacic's powerful algorithm to compute Liouvillian solutions to linear differential equations of order 2. Each chapter ends with a selection of exercise statements ranging in difficulty from the direct application of the theory to dealing with some topics that go beyond it. The reader will also find several illuminating examples. We have included a chapter with a list of further reading outlining the different directions in which differential Galois theory and related topics are being developed.

As guidance for teachers interested in using this book for a postgraduate course, we propose three possible courses, depending on the background and interests of their students.

(1) For students with limited or no knowledge of algebraic geometry who wish to understand Galois theory of linear differential equations in all its depth, a two–semester course can be given using the whole book.

(2) For students with good knowledge of algebraic geometry and algebraic groups, a one–semester course can be given based on Part 3 of the book using the first two parts as reference as needed.

(3) For students without a good knowledge of algebraic geometry and eager to learn differential Galois theory more quickly, a one–semester course can be given by developing the topics included in the following sections: 1.1, 3.1, 3.2, 3.3, 4.4 (skipping the references to Lie algebra), 4.6, and Part 3 (except the proof of Proposition 6.3.5, i.e. that the intermediate field of a Picard-Vessiot extension fixed by a normal closed subgroup of the differential Galois group is a Picard-Vessiot extension of the base field). This means introducing the concept of affine variety, defining the algebraic group and its properties considering only affine ones, determining the subgroups of $SL(2,\mathbb{C})$ assuming as a fact that a connected linear group of dimension less than or equal to 2 is solvable, and developing differential Galois theory (skipping the proof of Proposition 6.3.5).

Part 1

Algebraic Geometry

In Part 1, we introduce algebraic varieties and develop the related topics using an elementary approach. In the first chapter we define affine varieties as subsets of an affine space given by a finite set of polynomial equations. We see that affine varieties have a natural topology called Zariski topology. We introduce the concept of abstract affine variety to illustrate that giving an affine variety is equivalent to giving the ring of regular functions on each open set. We then study projective varieties and see how functions defined on a projective variety can be recovered by means of the open covering of the projective space by affine spaces.

In the second chapter we study algebraic varieties, which include affine and projective ones. We define morphism of algebraic varieties, the dimension of an algebraic variety, and the tangent space at a point. We analyze the dimension of the tangent space and define simple and singular points of a variety. We establish Chevalley's theorem and Zariski's main theorem which will be used in the construction of the quotient of an algebraic group by a closed subgroup.

For more details on algebraic geometry see [**Hu**], [**Kle**], and [**Sp**]. For the results of commutative algebra see [**A-M**], [**L**], and [**Ma**].

Unless otherwise specified, C will denote an algebraically closed field of characteristic 0.

Chapter 1

Affine and Projective Varieties

In this chapter we define an affine variety as the set of points of the affine space $\mathbb{A}^n$ over the field C which are common zeros of a finite set of polynomials in $C[X_1,\dots,X_n]$. An important result is Hilbert's Nullstellensatz which establishes a bijective correspondence between affine varieties of $\mathbb{A}^n$ and radical ideals of the polynomial ring $C[X_1,\dots,X_n]$. We define analogously projective varieties of the projective space $\mathbb{P}^n$ as the set of common zeros of a finite set of homogeneous polynomials, and we state a projective Nullstellensatz.

1.1. Affine varieties

Let $C[X_1,X_2,\dots,X_n]$ denote the ring of polynomials in the variables $X_1,X_2,\dots,X_n$ over C. The set $C^n = C\times\cdots\times C$ will be called *affine n-space* and denoted by $\mathbb{A}^n_C$ or just $\mathbb{A}^n$. We define an *affine variety* as the set of common zeros in $\mathbb{A}^n_C$ of a finite collection of polynomials in $C[X_1,\dots,X_n]$. To each ideal I of $C[X_1,\dots,X_n]$ we associate the set $\mathcal{V}(I)$ of its common zeros in $\mathbb{A}^n_C$. By Hilbert's basis theorem, the C-algebra $C[X_1,\dots,X_n]$ is Noetherian; hence each ideal of $C[X_1,\dots,X_n]$ has a finite set of generators. Therefore the set $\mathcal{V}(I)$ is an affine variety. To each subset $S\subset\mathbb{A}^n_C$ we associate the collection $\mathcal{I}(S)$ of all polynomials vanishing on S. It is clear that $\mathcal{I}(S)$ is an ideal and that we have inclusions $S\subset\mathcal{V}(\mathcal{I}(S))$, $I\subset\mathcal{I}(\mathcal{V}(I))$, which are not equalities in general.

Example 1.1.1. If $f\in C[X_1,X_2,\dots,X_n]\setminus C$, the affine variety $\mathcal{V}(f)$ is called a *hypersurface* of $\mathbb{A}^n_C$.

If $P = (x_1, \dots, x_n) \in \mathbb{A}^n_C$, $\{P\} = \mathcal{V}(X_1 - x_1, \dots, X_n - x_n)$ is an affine variety.

The following two propositions are easy to prove.

Proposition 1.1.2. *Let S, S_1, S_2 denote subsets of $\mathbb{A}^n_C$, I_1, I_2 denote ideals of $C[X_1, \dots, X_n]$. We have*

a) $S_1 \subset S_2 \Rightarrow \mathcal{I}(S_1) \supset \mathcal{I}(S_2)$,

b) $I_1 \subset I_2 \Rightarrow \mathcal{V}(I_1) \supset \mathcal{V}(I_2)$,

c) $\mathcal{I}(S) = C[X_1, X_2, \dots, X_n] \Leftrightarrow S = \emptyset$.

Proposition 1.1.3. *The correspondence $\mathcal{V}$ satisfies the following equalities:*

a) $\mathbb{A}^n_C = \mathcal{V}(0), \emptyset = \mathcal{V}(C[X_1, \dots, X_n])$,

b) If I and J are two ideals of $C[X_1, \dots, X_n]$, $\mathcal{V}(I) \cup \mathcal{V}(J) = \mathcal{V}(I \cap J)$,

c) If $\{I_\alpha\}$ is an arbitrary collection of ideals of $C[X_1, \dots, X_n]$, $\cap_\alpha \mathcal{V}(I_\alpha) = \mathcal{V}(\sum_\alpha I_\alpha)$.

We then have that affine varieties in $\mathbb{A}^n_C$ satisfy the axioms of closed sets in a topology. This topology is called *Zariski topology.* Hilbert's basis theorem implies the descending chain condition on closed sets and therefore the ascending chain condition on open sets. Hence $\mathbb{A}^n_C$ is a Noetherian topological space. This implies that it is quasicompact. However, the Hausdorff condition fails.

Example 1.1.4. For a point $P = (x_1, x_2, \dots, x_n) \in \mathbb{A}^n_C$, the ideal $\mathcal{I}(P) = (X_1 - x_1, X_2 - x_2, \dots, X_n - x_n)$ is maximal, as it is the kernel of the evaluation morphism

$$\begin{array}{rccc} v_P : & C[X_1, X_2, \dots X_n] & \to & C \\ & f & \mapsto & f(P). \end{array}$$

We recall that for an ideal I of a commutative ring A the *radical* $\sqrt{I}$ of I is defined by

$$\sqrt{I} := \{a \in A : a^r \in I \text{ for some } r \geq 1\}.$$

It is an ideal of A containing I. A *radical ideal* is an ideal which is equal to its radical. An ideal I of the ring A is radical if and only if the quotient ring A/I has no nonzero nilpotent elements. As examples of radical ideals, we have that a prime ideal is radical and ideals of the form $\mathcal{I}(S)$ for $S \subset \mathbb{A}^n_C$ are radical ideals of $C[X_1, \dots, X_n]$.

Example 1.1.5. The ideal $(X_1 X_2)$ is a radical ideal of $C[X_1, \dots, X_n]$ which is not prime. The ideal $(X^2 - 1)$ is a radical ideal of $C[X]$ which is not prime.

For an ideal I of $C[X_1, \dots, X_n]$, we can easily see the inclusion $\sqrt{I} \subset \mathcal{I}(\mathcal{V}(I))$. When the field C is algebraically closed, equality is given by the next theorem.

Theorem 1.1.6. (Hilbert's Nullstellensatz) *Let C be an algebraically closed field and let $A = C[X_1, \dots, X_n]$. Then the following hold:*

a) Every maximal ideal $\mathfrak{M}$ of A is of the form

$$\mathfrak{M} = (X_1 - x_1, X_2 - x_2, \dots, X_n - x_n) = \mathcal{I}(P),$$

for some point $P = (x_1, x_2, \dots, x_n)$ in $\mathbb{A}^n_C$.

b) If I is a proper ideal of A, then $\mathcal{V}(I) \neq \emptyset$.

c) If I is any ideal in A, then

$$\sqrt{I} = \mathcal{I}(\mathcal{V}(I)).$$

Remark 1.1.7. Point b) justifies the name of the theorem, namely "theorem on the zeros". To see the necessity of the condition C algebraically closed, we can consider the ideal $(X^2 + 1)$ in $\mathbb{R}[X]$.

For the proof of Hilbert's Nullstellensatz we shall use the following result, which is valuable on its own.

Proposition 1.1.8 (Noether's normalization lemma). *Let C be an arbitrary field, $R = C[x_1, \dots, x_n]$ a finitely generated C-algebra. Then there exist elements $y_1, \dots, y_r \in R$, with $r \leq n$, algebraically independent over C such that R is integral over $C[y_1, \dots, y_r]$.*

Proof. Let $\varphi : C[X_1, \dots, X_n] \to C[x_1, \dots, x_n]$ be the C-algebra morphism given by $\varphi(X_i) = x_i$, $1 \leq i \leq n$. Clearly, φ is an epimorphism. If it is an isomorphism, we just take $y_i := x_i$, $1 \leq i \leq n$. If not, let $f = \sum a_{i_1 \dots i_n} X_1^{i_1} \dots X_n^{i_n}$ be a nonzero polynomial in $\operatorname{Ker} \varphi$. We introduce an order relation in the set of monomials by defining $a_{i_1 \dots i_n} X_1^{i_1} \dots X_n^{i_n} < a_{i'_1 \dots i'_n} X_1^{i'_1} \dots X_n^{i'_n}$ if and only if $(i_1, \dots, i_n) < (i'_1, \dots, i'_n)$, with respect to the lexicographical order, i.e. for some $k \in \{1, \dots, n\}$, we have $i_l = i'_l$ if $l < k$ and $i_k < i'_k$. Let $a_{j_1 \dots j_n} X_1^{j_1} \dots X_n^{j_n}$ be the largest nonzero monomial in f. We can assume $a_{j_1 \dots j_n} = 1$. Let now d be an integer greater than all the exponents of the n variables appearing in the nonzero monomials of f. We consider the polynomial

$$h(X_1, \dots, X_n) := f(X_1 + X_n^{d^{n-1}}, X_2 + X_n^{d^{n-2}}, \dots, X_{n-1} + X_n^d, X_n).$$

The monomial $a_{i_1 \dots i_n} X_1^{i_1} \dots X_n^{i_n}$ in f becomes under the change of variables $a_{i_1 \dots i_n} X_n^{i_1 d^{n-1} + i_2 d^{n-2} + \dots + i_{n-1} d + i_n} +$ terms of lower degree in X_n; hence h is monic with respect to X_n and its leading term is X_n^c with $c = j_1d^{n-1} + j_2d^{n-2} + \cdots + j_{n-1}d + j_n$. The monic polynomial

$$g(X) := h(x_1 - x_n^{d^{n-1}}, x_2 - x_n^{d^{n-2}}, \ldots, x_{n-1} - x_n^d, X)$$

then satisfies $g(x_n) = f(x_1, \ldots, x_n) = 0$, which gives that the ring

$$C[x_1, x_2, \ldots, x_n] = C[x_1 - x_n^{d^{n-1}}, x_2 - x_n^{d^{n-2}}, \ldots, x_{n-1} - x_n^d, x_n]$$

is integral over the ring

$$R_1 := C[x_1 - x_n^{d^{n-1}}, x_2 - x_n^{d^{n-2}}, \ldots, x_{n-1} - x_n^d].$$

If the ring R_1 is isomorphic to the ring of polynomials $C[X_1, \ldots, X_{n-1}]$, then the proof is finished. Otherwise, by repeating the process we obtain a ring R_2 generated over C by $n-2$ elements over which R_1 is integral and, by transitivity of integral dependence, R as well . Since R is finitely generated over C, in a finite number of steps we obtain the result. The case in which $r = 0$, i.e. R is integral over C, is not excluded. □

Remark 1.1.9. Let us consider the ring R in Proposition 1.1.8 and let us assume that it is an integral domain. If we denote by K the fraction field of R, we have that the elements $y_1, \ldots, y_r$ form a transcendence basis of K over C; hence the transcendence degree of K over C is equal to r.

Corollary 1.1.10. *Let $S \subset R$ be two finitely generated domains over C. Then there exist elements $f \in S$ and $x_1, \ldots, x_r \in R$, algebraically independent over S_f such that R_f is integral over $S_f[x_1, \ldots, x_r]$.*

Proof. Let $K \subset L$ be the fields of fractions of S and R. Denote by R' the localization of R with respect to the multiplicative system S^* of the nonzero elements of S. We apply Proposition 1.1.8 to the finitely generated K-algebra R'. There exist elements $x_1, \ldots, x_r$ in R' algebraically independent over K such that R' is integral over $K[x_1, \ldots, x_r]$. It is clear that the elements x_i can be chosen in R as the denominators appearing are invertible in K. Now R is finitely generated over S and each generator satisfies an integral dependence relation over $K[x_1, \ldots, x_r]$. By choosing a common denominator f for the coefficients of all these relations, we obtain that R_f is integral over $S_f[x_1, \ldots, x_r]$. □

For the proof of the Hilbert's Nullstellensatz, we use the following proposition, sometimes called "weak Nullstellensatz".

Proposition 1.1.11. *Let C be an arbitrary field, $R = C[x_1, \ldots, x_n]$ a finitely generated C-algebra. If R is a field, then it is algebraic over C.*

Proof. By Noether's normalization Lemma 1.1.8, there exist elements $y_1, \ldots, y_r \in R$, with $r \leq n$, algebraically independent over C such that R is integral over $A := C[y_1, \ldots, y_d]$, hence a finite A-algebra. This implies that A is a field. Indeed, a nonzero element α in A has an inverse α^{-1} in R. Writing down an integral dependence relation for α^{-1} over A, $\alpha^{-n} + a_1 \alpha^{-n+1} + \cdots + a_{n-1} \alpha^{-1} + a_n = 0$, and multiplying it by α^{n-1}, we obtain $\alpha^{-1} = -(a_1 + \cdots + a_{n-1} \alpha^{n-2} + a_n \alpha^{n-1}) \in A$. But A can only be a field for $d = 0$, so R is a finite extension of C, hence algebraic over C. □

Proof of Hilbert's Nullstellensatz.

a) Let $\mathfrak{M}$ be a maximal ideal in A. Then we have that $K := A/\mathfrak{M}$ is a field, which is generated over C by the residue classes $X_i \bmod \mathfrak{M}$. By Proposition 1.1.11, K is algebraic over C and, as C is algebraically closed, the natural morphism of C-algebras

$$\varphi : C \hookrightarrow C[X_1, X_2, \ldots X_n] \stackrel{\pi}{\to} C[X_1, X_2, \ldots, X_n]/\mathfrak{M} = K$$

is an isomorphism between C and K. Let $x_i := \varphi^{-1}(X_i \bmod \mathfrak{M})$, $1 \leq i \leq n$. Then $X_i - x_i \in \operatorname{Ker} \pi = \mathfrak{M}$ and so, $(X_1 - x_1, X_2 - x_2, \ldots, X_n - x_n) \subset \mathfrak{M}$. As $(X_1 - x_1, X_2 - x_2, \ldots, X_n - x_n)$ is a maximal ideal, we have equality.

b) Let $I \subsetneq A$. There exists a maximal ideal $\mathfrak{M}$ of A such that $I \subset \mathfrak{M}$. From a) we have $\mathfrak{M} = \mathcal{I}(P)$ for some point $P \in \mathbb{A}^n_C$, so $\{P\} \subset \mathcal{V}(\mathcal{I}(P)) \subset \mathcal{V}(I)$; hence $\mathcal{V}(I)$ is not empty.

c) For an ideal I of A we want to prove that if f is an element in $\mathcal{I}(\mathcal{V}(I))$, then f^r belongs to I for some integer r. We shall use *Rabinowitsch's trick*, which consists of adding a variable T and considering the natural inclusion $C[X_1, X_2, \ldots, X_n] \subset C[X_1, X_2, \ldots, X_n, T]$ and the ideal $J = (I, Tf - 1)$ of $C[X_1, X_2, \ldots, X_n, T]$. We clearly have

$$\mathcal{V}(J) = \{(x_1, x_2, \ldots, x_n, y) = (P, y) \in \mathbb{A}^{n+1}_C : P \in \mathcal{V}(I) \text{ and } \ yf(P) = 1\}.$$

Projection onto the first n coordinates maps $\mathcal{V}(J)$ to the subset of $\mathcal{V}(I)$ of points P with $f(P) \neq 0$. But f belongs to $\mathcal{I}(\mathcal{V}(I))$, so $\mathcal{V}(J) = \emptyset$. By b), we then have $J = C[X_1, X_2, \ldots, X_n, T]$. In particular, $1 \in J$, so we can write

$$(1.1) \qquad 1 = \sum_{i=1}^{m} g_i f_i + g_0(Tf - 1),$$

for some $g_i \in C[X_1, X_2, \ldots, X_n, T]$, $f_i \in I$. Let T^r be the highest power of T appearing in the polynomials g_i, for $0 \leq i \leq m$. Multiplying (1.1) by f^r gives

$$f^r = \sum_{i=1}^{m} G_i f_i + G_0(Tf - 1),$$

where the $G_i = f^r g_i$ are polynomials in $X_1, \ldots, X_n, Tf$. Considering this last equality modulo $Tf - 1$, we obtain the relationship

$$f^r \equiv \sum_{i=1}^{m} h_i f_i \mod(Tf - 1),$$

where $h_i(X_1, \ldots, X_n) := G_i(X_1, \ldots, X_n, 1), 1 \leq i \leq m$. Now the natural morphism

$$C[X_1, X_2, \ldots, X_n] \hookrightarrow C[X_1, X_2, \ldots, X_n, T] \twoheadrightarrow C[X_1, X_2, \ldots, X_n, T]/(Tf - 1)$$

is injective. So we have the equality

$$f^r = \sum_{i=1}^{m} h_i f_i$$

in $C[X_1, \ldots, X_n]$. Thus $f^r \in I$. □

Remark 1.1.12. In the proof of Hilbert's Nullstellensatz, the hypothesis C algebraically closed was only used to prove a). Then b) was proved assuming a) and c) was proved assuming b). For any field C it can be proved that the three statements are equivalent. Indeed assuming c), if $\mathfrak{M}$ is a maximal ideal of $C[X_1, X_2, \ldots, X_n]$, we have $\mathcal{I}(\mathcal{V}(\mathfrak{M})) = \sqrt{\mathfrak{M}} = \mathfrak{M} \subsetneq C[X_1, X_2, \ldots, X_n]$; hence $\mathcal{V}(\mathfrak{M}) \neq \emptyset$. If $P \in \mathcal{V}(\mathfrak{M})$, then $\mathfrak{M} \subset \mathcal{I}(P)$, and as $\mathfrak{M}$ is maximal, $\mathfrak{M} = \mathcal{I}(P)$.

As a consequence of Hilbert's Nullstellensatz, we have that $\mathcal{V}$ and $\mathcal{I}$ set a bijective correspondence between the collection of all radical ideals of $C[X_1, \ldots, X_n]$ and the collection of all affine varieties of $\mathbb{A}^n_C$ which inverts inclusion.

Recall that a nonempty topological space X is said to be reducible if it can be written as a union of two closed proper subsets. It is *irreducible* if it is not reducible, or equivalently, if all nonempty open subsets of X are dense. A subset of a topological space is reducible (resp. irreducible) if it is so as a topological space with the induced topology. Recall as well that a Noetherian topological space can be written as a union of its irreducible components, i.e. its finitely many maximal irreducible subsets. If a subset is irreducible, its closure is also; so irreducible components are closed.

For closed subsets in $\mathbb{A}^n_C$ irreducibility is characterized in terms of the corresponding ideal by the following proposition.

Proposition 1.1.13. *A closed set V in $\mathbb{A}^n_C$ is irreducible if and only if its ideal $\mathcal{I}(V)$ is prime. In particular, $\mathbb{A}^n_C$ itself is irreducible.*

Proof. Write $I = \mathcal{I}(V)$. Suppose that V is irreducible and let $f_1, f_2 \in C[X_1, \ldots, X_n]$ such that $f_1 f_2 \in I$. Then each $P \in V$ is a zero of f_1 or f_2; hence $V \subset \mathcal{V}(I_1) \cup \mathcal{V}(I_2)$, for I_i the ideal generated by $f_i, i = 1, 2$. Since V is irreducible, it must be contained within one of these two sets, i.e. $f_1 \in I$ or $f_2 \in I$, and I is prime.

Reciprocally, assume that I is prime but $V = V_1 \cup V_2$, with V_1, V_2 closed in V. If none of the V_i's covers V, we can find $f_i \in \mathcal{I}(V_i)$ but $f_i \notin I$, $i = 1, 2$. But $f_1 f_2$ vanish on V, so $f_1 f_2 \in I$, contradicting that I is prime. □

Example 1.1.14. As $C[X_1, X_2, \ldots, X_n]$ is a unique factorization domain, for $f \in C[X_1, \ldots, X_n] \setminus C$, the irreducible components of the hypersurface $\mathcal{V}(f)$ in $\mathbb{A}^n$ are just the hypersurfaces defined as the zero sets of the irreducible factors of f.

Example 1.1.15. For the closed set $\mathcal{V}(X_1 X_2) \subset \mathbb{A}^n$, $\mathcal{V}(X_1 X_2) = \mathcal{V}(X_1) \cup \mathcal{V}(X_2)$ is the decomposition as the union of its irreducible components which are coordinate hyperplanes. For the closet set $\mathcal{V}(X^2 - 1) \subset \mathbb{A}^2_C$, $\mathcal{V}(X^2 - 1) = \mathcal{V}(X-1) \cup \mathcal{V}(X+1)$ is the descomposition as the union of its irreducible components which are points.

From Hilbert's Nullstellensatz and Proposition 1.1.13, we obtain that $\mathcal{V}$ and $\mathcal{I}$ set the following bijective correspondences.

$$\begin{array}{lcl}
\{\text{radical ideals of } C[X_1, X_2, \ldots, X_n]\} & \leftrightarrow & \{\text{closed sets in } \mathbb{A}^n_C\}, \\
\{\text{prime ideals of } C[X_1, X_2, \ldots, X_n]\} & \leftrightarrow & \{\text{irreducible closed sets in } \mathbb{A}^n_C\}, \\
\{\text{maximal ideals of } C[X_1, X_2, \ldots, X_n]\} & \leftrightarrow & \{\text{points in } \mathbb{A}^n_C\}.
\end{array}$$

A *principal open set* of $\mathbb{A}^n_C$ is the set of non-zeros of a single polynomial. We note that principal open sets are a basis of the Zariski topology. The closure in the Zariski topology of a principal open set is the whole affine space. Hence, as $\mathbb{A}^n_C$ is irreducible, we obtain that principal open sets are irreducible.

If V is closed in $\mathbb{A}^n_C$, each polynomial $f(X) \in C[X_1, \dots, X_n]$ defines a C-valued function on V. But different polynomials may define the same function. It is clear that we have a 1-1 correspondence between the distinct polynomial functions on V and the residue class ring $C[X_1, \dots, X_n]/\mathcal{I}(V)$. We denote this ring by $C[V]$ and call it the *coordinate ring* of V. It is a finitely generated algebra over C and is reduced (i.e. without nonzero nilpotent elements) because $\mathcal{I}(V)$ is a radical ideal.

Remark 1.1.16. If V is an affine variety in $\mathbb{A}^n_C$, we can consider in V the Zariski topology induced by the topology in $\mathbb{A}^n_C$. The closed sets in this topology can be defined as $\mathcal{V}(I) := \{P \in V : f(P) = 0, \ \forall f \in I\}$ for an ideal I of $C[V]$.

If V is irreducible, equivalently if $\mathcal{I}(V)$ is a prime ideal, $C[V]$ is an integral domain. We can then consider its field of fractions $C(V)$, which is called *function field* of V. Elements $f \in C(V)$ are called *rational functions* on V. Any rational function can be written $f = g/h$, with $g, h \in C[V]$. In general, this representation is not unique. We can only give f a well-defined value at a point P if there is a representation $f = g/h$, with $h(P) \neq 0$. In this case we say that the rational function f is *regular at* P. The *domain of definition* of f is defined to be the set

$$\mathrm{dom}(f) = \{P \in V : f \text{ is regular at } P\}.$$

Example 1.1.17. We consider $V := \mathcal{V}(Y^2 - X^3 + X) \subset \mathbb{A}^2_{\mathbb{C}}$ and $P = (0,0) \in V$. The function X/Y is regular at P as it can be written as $Y/(X^2 - 1)$ in $\mathbb{C}(V)$.

Proposition 1.1.18. *Let V be an irreducible variety. For a rational function $f \in C(V)$, the following statements hold*

a) $\mathrm{dom}(f)$ *is open and dense in V.*

b) $\mathrm{dom}(f) = V \Leftrightarrow f \in C[V]$.

c) If $h \in C[V]$ and $V_h := \{P \in V : h(P) \neq 0\}$, then $\mathrm{dom}(f) \supset V_h \Leftrightarrow f \in C[V][1/h]$.

Proof. a) For $f \in C(V)$, we consider the ideal of denominators

$$J_f = \{h \in C[V] : hf \in C[V]\} \subset C[V].$$

Then $\operatorname{dom}(f) = \{P \in V : h(P) \neq 0 \text{ for some } h \in J_f\}$; hence its complement $V \setminus \operatorname{dom}(f) = \mathcal{V}(J_f)$ is closed. As $\operatorname{dom}(f)$ is an open subset of the irreducible closed subset V, it is dense in V.

b) $\operatorname{dom}(f) = V \Leftrightarrow \mathcal{V}(J_f) = \emptyset$. By Hilbert's Nullstellensatz 1.1.6, this last equality implies $1 \in J_f$ and so $f \in C[V]$.

c) We have $\operatorname{dom}(f) \supset V_h$ if and only if h vanishes on $\mathcal{V}(J_f)$. By Hilbert's Nullstellensatz, this is equivalent to $h^r \in J_f$ for some $r \geq 1$. This means that $f = g/h^r \in C[V][1/h]$. □

Part b) of Proposition 1.1.18 says that the polynomial functions are precisely the rational functions that are "everywhere regular".

If U is an open subset of an irreducible variety V, a rational function f in $C(V)$ is *regular on* U if it is regular at each point of U. The set of rational functions of $C(V)$ which are regular on U is a subring of $C(V)$. We denote it by $\mathcal{O}(U)$.

The *local ring of* V *at a point* $P \in V$ is the ring

$$\mathcal{O}_P := \{f \in C(V) : f \text{ is regular at } P\}.$$

It is isomorphic to the ring $C[V]_{\mathfrak{M}_P}$ obtained by localizing the ring $C[V]$ at the maximal ideal $\mathfrak{M}_P = \{f \in C[V] : f(P) = 0\}$. This is indeed a local ring, i.e. it has a unique maximal ideal, namely $\mathfrak{M}_P C[V]_{\mathfrak{M}_P}$.

Remark 1.1.19. If V is an irreducible affine variety, then

$$C[V] = \bigcap_{P \in V} \mathcal{O}_P.$$

Indeed, as $C[V]$ is an integral domain, we can apply **[Ma]** Lemma 2, p.8.

If V is an arbitrary affine variety, U an open subset of V, a function $f : U \to C$ is regular at a point x in V if there exists $g, h \in C[V]$ and an open subset U_0 of U containing x such that for all $y \in U_0$, $h(y) \neq 0$ and $f(y) = g(y)/h(y)$. A function $f : U \to C$ is regular in an open subset U' of U if it is regular at each point of U'.

Let us observe that a principal open set can be seen as an affine variety. If $V_f = \{x \in \mathbb{A}^n_C : f(x) \neq 0\}$, for some $f \in C[X_1, \ldots, X_n]$, the points of V_f are in 1-1 correspondence with the points of the closed set of $\mathbb{A}^{n+1}_C$ $\{(x_1, \ldots, x_n, x_{n+1}) : f(x_1, \ldots, x_n)\, x_{n+1} - 1 = 0\}$. Hence V_f has an affine variety structure and its coordinate ring is $C[V_f] = C[X_1, \ldots, X_n, 1/f]$, i.e. the ring $C[X_1, \ldots, X_n]$ localized in the multiplicative system of the powers of f.

More generally, for V an affine variety, $f \in C[V]$, the algebra of regular functions on the principal open set $V_f := \{x \in V : f(x) \neq 0\}$ is the algebra $C[V]_f$, i.e. the algebra $C[V]$ localized in the multiplicative system of the powers of f. We note that arbitrary open sets of an affine variety cannot be given the structure of an affine variety. (See Exercise 11.)

Now let $V \subset \mathbb{A}^n_C$, $W \subset \mathbb{A}^m_C$ be arbitrary affine varieties. A map $\varphi : V \to W$ is a *morphism* of affine varieties if for $x = (x_1, \ldots, x_n) \in V$, $\varphi(x_1, \ldots, x_n) = (\varphi_1(x), \ldots, \varphi_m(x))$, for some $\varphi_i \in C[V]$. A morphism $\varphi : V \to W$ is continuous for the Zariski topologies involved. Indeed if $Z \subset W$ is the set of zeros of polynomial functions f_i on W, then $\varphi^{-1}(Z)$ is the set of zeros of the polynomial functions $f_i \circ \varphi$ on V. With a morphism $\varphi : V \to W$, a C-algebra morphism $\varphi^* : C[W] \to C[V]$ is associated, defined by $\varphi^*(f) = f \circ \varphi$. Note that if $y_1, \ldots, y_m$ are the coordinate functions on W, we have $\varphi_j = \varphi^*(y_j)$; hence φ is recovered from φ^*. If X is a third affine variety and $\psi : W \to X$ a morphism, we clearly have $(\psi \circ \varphi)^* = \varphi^* \circ \psi^*$.

Proposition 1.1.20. *If $\varphi : V \to W$ is a morphism of affine varieties for which $\varphi(V)$ is dense in W, then φ^* is injective.*

Proof. Let $f, g \in C[W]$ such that $\varphi^*(f) = \varphi^*(g)$. We consider the subset $\{y \in W : f(y) = g(y)\}$ of W. It is clearly closed. On the other hand it contains $\varphi(\{x \in V : f(\varphi(x)) = g(\varphi(x))\}) = \varphi(V)$; hence it is dense in W. So it is equal to W. □

The morphism $\varphi : V \to W$ is an *isomorphism* if there exists a morphism $\psi : W \to V$ such that $\psi \circ \varphi = Id_V$ and $\varphi \circ \psi = Id_W$, or equivalently $\varphi^* : C[W] \to C[V]$ is an isomorphism of C-algebras (with its inverse being ψ^*). We say that the varieties V, W defined over the same field C are *isomorphic* if there exists an isomorphism $\varphi : V \to W$.

We will often need to consider maps on an irreducible affine variety V which are not everywhere defined, so we introduce the following concept.

Definition 1.1.21. a) If V is an irreducible affine variety, a *rational map* $\varphi : V \to \mathbb{A}^n_C$ is an n-tuple $(\varphi_1, \ldots, \varphi_n)$ of rational functions $\varphi_1, \ldots, \varphi_n \in C(V)$. The map φ is called *regular* at a point P of V if all φ_i are regular at P. The *domain of definition* $\operatorname{dom}(\varphi)$ is the set of all regular points of φ, i.e. $\operatorname{dom}(\varphi) = \bigcap_{i=1}^n \operatorname{dom}(\varphi_i)$.

b) For an affine variety $W \subset \mathbb{A}^n_C$, a *rational map* $\varphi : V \to W$ is an n-tuple $(\varphi_1, \ldots, \varphi_n)$ of rational functions $\varphi_1, \ldots, \varphi_n \in C(V)$ such that $\varphi(P) := (\varphi_1(P), \ldots, \varphi_n(P)) \in W$ for all $P \in \text{dom}(\varphi)$.

A rational map $\varphi : V \to W$ induces a C-algebra morphism $\varphi^* : C[W] \to C(V)$ given by $g \mapsto g \circ \varphi$. To determine when it is possible to extend this morphism to $C(W)$, in the case when W is also irreducible, we make the following definition.

Definition 1.1.22. A rational map $\varphi : V \to W$ is called *dominant* if $\varphi(\text{dom}(\varphi))$ is a Zariski dense subset of W.

Proposition 1.1.23. *For irreducible affine varieties V and W, the following hold.*

a) Every dominant rational map $\varphi : V \to W$ induces a C-linear morphism $\varphi^* : C(W) \to C(V)$.

b) If $f : C(W) \to C(V)$ is a field homomorphism which is C-linear, then there exists a unique dominant rational map $\varphi : V \to W$ with $f = \varphi^*$.

c) If $\varphi : V \to W$ and $\psi : W \to X$ are dominant, then $\psi \circ \varphi : V \to X$ is also dominant and $(\psi \circ \varphi)^* = \varphi^* \circ \psi^*$.

Proof. a) If φ is defined by $\varphi_1, \ldots, \varphi_n$, with $\varphi_i \in C(V)$, then for $g \in C[W]$, $g(\varphi_1, \ldots, \varphi_n) \in C(V)$. Hence φ induces a C-algebra morphism $\varphi^* : C[W] \to C(V)$. Now if $\varphi^*(g) = 0$, then g vanishes on $\varphi(\text{dom}(\varphi))$, which is dense in W, so $g = 0$, hence φ^* is injective, so it extends to $C(W)$.

b) If $W \subset \mathbb{A}^n$, then the restrictions $x_1, x_2, \ldots, x_n$ of the coordinate functions to W, generate $C(W)$. Let $\varphi_i := f(x_i) \in C(V)$ and $\varphi := (\varphi_1, \ldots, \varphi_n)$. Then φ defines a rational map from V to W and $f = \varphi^*$ by construction. Now $\varphi^*_{|C[W]} = f_{|C[W]}$ is injective, so φ is dominant, as otherwise there will be nonzero elements in $C[W]$ vanishing at $\varphi(\text{dom}(\varphi))$, hence in $\operatorname{Ker} \varphi^*$.

c) As $\varphi(\text{dom}(\varphi)) \cap \text{dom}\, \psi \neq \emptyset$, the composition makes sense and both statements are clear. □

Definition 1.1.24. Let V, W be irreducible affine varieties. A rational map $\varphi : V \to W$ is called *birational* (or a *birational equivalence*) if there is a rational map $\psi : W \to V$ with $\varphi \circ \psi = Id_{\text{dom}(\psi)}$ and $\psi \circ \varphi = Id_{\text{dom}(\varphi)}$.

Definition 1.1.25. Two irreducible varieties V and W are said to be *birationally equivalent* if there is a birational equivalence $\varphi : V \to W$.

Proposition 1.1.26. *Let V, W be irreducible affine varieties. For a rational map $\varphi : V \to W$, the following statements are equivalent.*

a) φ is birational.

b) φ is dominant and $\varphi^* : C(W) \to C(V)$ is an isomorphism.

c) *There are nonempty open sets* $V_0 \subset V$ *and* $W_0 \subset W$ *such that the restriction* $\varphi_{|V_0} : V_0 \to W_0$ *is an isomorphism.*

Proof. a) $\Rightarrow$ b) The rational map $\psi : W \to V$ such that $\varphi \circ \psi = Id_W$ and $\psi \circ \varphi = Id_V$ is regular in a dense open set of W; hence φ is dominant. As $\varphi(V)$ is dense in W, φ^* is injective. Analogously, ψ^* is injective.

b) $\Rightarrow$ a) We define $\psi : W \to V$ by taking $\psi := (\varphi^{*-1}(x_i))_i$ for x_i the restriction to V of the coordinate functions. By construction we have $\varphi \circ \psi = Id_W$ and $\psi \circ \varphi = Id_V$.

a) $\Rightarrow$ c) Let $\varphi = (\varphi_1, \dots, \varphi_n)$ and $\varphi_i = f_i/F$ for $f_i, F \in C[V]$. Then $\varphi_i \in C[V_F]$, for V_F the principal open set defined by the nonvanishing of F. Analogously, if $\psi = (\psi_1, \dots, \psi_n)$ and $\psi_i = g_i/G$ for $g_i, G \in C[W]$, then $\psi_i \in C[W_G]$. By taking $V_0 = V_F \cap \psi(\operatorname{dom} \psi)$ and $W_0 = W_G \cap \varphi(\operatorname{dom} \varphi)$, we obtain the result.

c) $\Rightarrow$ b) φ is dominant as $W_0 \subset \varphi(\operatorname{dom} \varphi)$ and φ^* is an isomorphism from $C(W) = C(W_0)$ onto $C(V) = C(V_0)$. □

We now prove that every irreducible affine variety is birationally equivalent to a hypersurface in some affine space. This fact will be used to determine the dimension of the tangent spaces of a variety.

Proposition 1.1.27. *Every irreducible affine variety* V *is birationally equivalent to a hypersurface in some affine space* $\mathbb{A}^n$.

Proof. The field $C(V)$ is finitely generated over C. Then by Proposition 1.1.8, there exist elements $x_1, \dots, x_d$ in $C(V)$, algebraically independent over C, such that $C(V)$ is algebraic over $C(x_1, \dots, x_d)$. Since we are assuming $\operatorname{char} C = 0$, we can apply the primitive element theorem and obtain $C(V) = C(x_1, \dots, x_d)(x_{d+1})$ for some element x_{d+1} algebraic over $C(x_1, \dots, x_d)$. We then have an algebraic dependence relation $f(x_1, \dots, x_d, x_{d+1}) = 0$, with $f \in C[X_1, \dots, X_{d+1}]$. Let W be the hypersurface in $\mathbb{A}^n$, with $n = d+1$, defined by the vanishing of the polynomial f. Then by construction $C(W) \simeq C(V)$, so V and W are birationally equivalent by Proposition 1.1.26. □

We shall now introduce the notion of dimension of an affine variety. If X is a noetherian topological space, we define the *dimension* of X to be the supremum of all integers n such that there exists a chain $Z_0 \subset Z_1 \subset \cdots \subset Z_n$ of distinct irreducible closed subsets of X. We define the dimension of an affine variety to be its dimension as a topological space. Clearly the dimension of an affine variety is the maximum of the dimensions of its irreducible components. For a ring A, we define the *Krull dimension* $\dim A$ of A to be the supremum of all integers n such that there exists a chain $P_0 \subset$

$P_1 \subset \cdots \subset P_n$ of distinct prime ideals of A. If $V \subset \mathbb{A}_C^n$ is an affine variety, by Proposition 1.1.13, irreducible closed subsets of V correspond to prime ideals of $C[X_1, \ldots, X_n]$ containing $\mathcal{I}(V)$ and these in turn correspond to prime ideals of $C[V]$. Hence the dimension of V is equal to the Krull dimension of its coordinate ring $C[V]$. We now recall that $\dim C[X_1, X_2, \ldots, X_n] = n$ ([**Ma**] § 14, Theorem 22) which, by the preceding, implies $\dim \mathbb{A}_C^n = n$. We recall as well that if a noetherian ring R is integral over a noetherian subring S, then $\dim S = \dim R$ ([**Ma**] § 13, Theorem 20). Now in the situation of Noether's normalization lemma (Proposition 1.1.8) and with the same notations there, we have $\dim R = r$. Hence the dimension of a finitely generated integral domain R over C is equal to the transcendence degree of its fraction field over C. (See Remark 1.1.9.) We then obtain that if V is irreducible, the dimension of V is equal to the Krull dimension of $C[V]$ and equal to the transcendence degree $\operatorname{trdeg}[C(V) : C]$ of the function field $C(V)$ of V over C.

We now give a geometric interpretation of Noether's normalization lemma. Let $V \subset \mathbb{A}^n$ be an affine irreducible variety. We consider the ring $C[V] = C[X_1, \ldots, X_n]/\mathcal{I}(V)$ and denote by x_i the image of X_i in $C[V]$, $1 \leq i \leq n$. Then $C[V] = C[x_1, \ldots, x_n]$ is a finitely generated integral domain over C. By Proposition 1.1.8, there exist elements $y_1, \ldots, y_r \in C[V]$, with $r \leq n$, algebraically independent over C such that $C[V]$ is integral over $C[y_1, \ldots, y_r]$. The elements y_i lift to elements $\widetilde{y_i} \in C[X_1, \ldots, X_n]$, which define a morphism $\psi = (\widetilde{y_1}, \ldots, \widetilde{y_r}) : \mathbb{A}^n \to \mathbb{A}^r$. The restriction φ of ψ to V is independent of the choice of the $\widetilde{y_i}$ as $\widetilde{y_i} \mod \mathcal{I}(V) = y_i$. We will now show that the fibers of φ are finite and nonempty.

Proposition 1.1.28. *Let $\varphi : V \to \mathbb{A}^r$ be defined as above. For each $P \in \mathbb{A}^r$, the fiber $\varphi^{-1}(P)$ is finite and nonempty.*

Proof. We first prove that $\varphi^{-1}(P)$ is finite. As $C[V]$ is integral over the ring $C[y_1, \ldots, y_r]$, there exist an integer N and polynomials f_j^i, $0 \leq j \leq N-1$, $1 \leq i \leq n$ such that

$$x_i^N + f_{N-1}^i(y_1, \ldots, y_r)\, x_i^{N-1} + \cdots + f_1^i(y_1, \ldots, y_r)\, x_i + f_0^i(y_1, \ldots, y_r) = 0, \quad 1 \leq i \leq n.$$

So, we have

$$X_i^N + f_{N-1}^i(\widetilde{y_1}, \ldots, \widetilde{y_r})\, X_i^{N-1} + \cdots + f_1^i(\widetilde{y_1}, \ldots, \widetilde{y_r})\, X_i + f_0^i(\widetilde{y_1}, \ldots, \widetilde{y_r}) =: g_i(X_1, \ldots, X_n) \in \mathcal{I}(V).$$

If $Q = (q_1, \dots, q_n) \in V$, we have $g_i(q_1, \dots, q_n) = 0$ and so q_i is a solution of the equation $f_i(q_i) = 0$, where

$$f_i(X) := X^N + f^i_{N-1}(y_1, \dots, y_r)\, X^{N-1} + \dots + f^i_1(y_1, \dots, y_r)\, X + f^i_0(y_1, \dots, y_r).$$

Now, as V is irreducible, we can consider its function field $C(V)$ and see $f_i(X)$ as a polynomial in $C(V)[X]$. Then each f_i has a finite number of roots. Thus for any point $P = (y_1, \dots, y_r) \in \mathbb{A}^r$ we have only finitely many points $Q \in V$ with $\varphi(Q) = P$.

To show that $\varphi^{-1}(P)$ is always nonempty, it is enough to show that for every point $P = (p_1, \dots, p_r) \in \mathbb{A}^r$, we have

$$(1.2) \qquad I_P := \mathcal{I}(V) + (y_1 - p_1, \dots, y_r - p_r) \neq C[X_1, \dots, X_n]$$

as, by Hilbert's Nullstellensatz, this will imply $\varphi^{-1}(P) = \mathcal{V}(I_P) \neq \emptyset$. Assertion (1.2) is equivalent to $(y_1 - p_1, \dots, y_r - p_r) \neq C[x_1, \dots, x_n]$. Since $(y_1 - p_1, \dots, y_r - p_r)$ is a maximal ideal in $C[y_1, \dots, y_r]$, in particular a proper ideal, we can apply Nakayama's lemma (see [**Ma**] p. 11) to the $C[y_1, \dots, y_r]$-finite algebra $C[V]$ and obtain $(y_1 - p_1, \dots, y_r - p_r) \neq C[x_1, \dots, x_n]$. □

The preceding proposition gives that an affine irreducible variety V of dimension r can be seen as a finite covering of the affine space $\mathbb{A}^r$.

We now consider extension of scalars for affine varieties. Let $V \subset \mathbb{A}^n_C$ be an affine variety and L an algebraically closed field containing C. We shall denote by V_L the affine variety contained in $\mathbb{A}^n_L$ defined by $V_L = \mathcal{V}(I_L)$ for $I_L = \mathcal{I}(V)L[X_1, \dots, X_n]$. We call V_L the variety obtained from V by extending scalars to L. The coordinate ring of V_L is $L[V] = L \otimes C[V]$. It is clear that if V, W are affine varieties defined over C, we have $V \simeq W \Rightarrow V_L \simeq W_L$. The next proposition gives the converse of this implication. For its proof we are following a suggestion of Jakub Byszewski.

Proposition 1.1.29. *Let K, L be algebraically closed fields with $K \subset L$. Let V, W be affine algebraic varieties defined over K. Let V_L, W_L be the varieties obtained from V, W by extending scalars to L. If V_L and W_L are isomorphic, then V and W are isomorphic.*

Proof. Let $A = K[V], B = K[W]$. These are finitely generated K-algebras. Let us write $A = K[x_1, \dots, x_n], B = K[y_1, \dots, y_m]$. The isomorphism between V_L and W_L induces an L-algebra isomorphism

$$f : A \otimes_K L = L[x_1, \dots, x_n] \xrightarrow{\sim} B \otimes L = L[y_1, \dots, y_m].$$

Then $f(x_i)$ is a polynomial in $y_1, \ldots, y_m$ with coefficients in L, $1 \leq i \leq n$, $f^{-1}(y_j)$ is a polynomial in $x_1, \ldots, x_n$ with coefficients in L, $1 \leq j \leq m$. Let $S \subset L$ be the K-algebra generated by the coefficients of all these polynomials. Then S is a finitely generated K-algebra and we have an isomorphism

$$A \otimes_K S \overset{\sim}{\to} B \otimes_K S. \tag{1.3}$$

As K is algebraically closed, using Proposition 1.1.11, we have either $S = K$, in which case we obtain $A \simeq A \otimes_K K \simeq B \otimes_K K \simeq B$; hence $V \simeq W$ as wanted or S is not a field. In this latter case, let $\mathfrak{M}$ be a maximal ideal in S. As S is a finitely generated K-algebra and K is algebraically closed, we have $S/\mathfrak{M} \simeq K$ (by Hilbert Nullstellensatz 1.1.6). Tensoring the isomorphism (1.3) by $S/\mathfrak{M}$ over S, we obtain

$$\begin{array}{ccc} A \otimes_K S \otimes_S S/\mathfrak{M} & \overset{\sim}{\to} & B \otimes_K S \otimes_S S/\mathfrak{M} \\ \downarrow\wr & & \downarrow\wr \\ A \otimes_K K = A & \overset{\sim}{\to} & B \otimes_K K = B. \end{array}$$

Hence $V \simeq W$. □

1.2. Abstract affine varieties

So far we have considered affine varieties as closed subsets of affine spaces. We shall now see that they can be defined in an intrinsic way (i.e. not depending on an embedding in an ambient space) as topological spaces endowed with a sheaf of functions satisfying the properties of the regular functions.

Definition 1.2.1. A *sheaf of functions* on a topological space X is a function $\mathcal{F}$ which assigns to every nonempty open subset $U \subset X$ a C-algebra $\mathcal{F}(U)$ of C-valued functions on U such that the following two conditions hold:

a) If $U \subset U'$ are two nonempty open subsets of X and $f \in \mathcal{F}(U')$, then the restriction $f_{|U}$ belongs to $\mathcal{F}(U)$.

b) Given a family of open sets $U_i, i \in I$, covering U and functions $f_i \in \mathcal{F}(U_i)$ for each $i \in I$, such that f_i and f_j agree on $U_i \cap U_j$, for each pair of indices i, j, there exists a function $f \in \mathcal{F}(U)$ whose restriction to each U_i equals f_i.

Definition 1.2.2. A topological space X together with a sheaf of functions $\mathcal{O}_X$ is called a *geometric space*. We refer to $\mathcal{O}_X$ as the *structure sheaf* of the geometric space X.

Definition 1.2.3. Let $(X, \mathcal{O}_X)$ and $(Y, \mathcal{O}_Y)$ be geometric spaces. A *morphism*

$$\varphi : (X, \mathcal{O}_X) \to (Y, \mathcal{O}_Y)$$

is a continuous map $\varphi : X \to Y$ such that for every open subset U of Y and every $f \in \mathcal{O}_Y(U)$, the function $\varphi^*(f) := f \circ \varphi$ belongs to $\mathcal{O}_X(\varphi^{-1}(U))$.

Example 1.2.4. Let X be an affine variety. To each nonempty open set $U \subset X$ we assign the ring $\mathcal{O}_X(U)$ of regular functions on U. Then $(X, \mathcal{O}_X)$ is a geometric space. Moreover the two notions of morphism agree.

Let $(X, \mathcal{O}_X)$ be a geometric space and let Z be a subset of X with the induced topology. We can make Z into a geometric space by defining $\mathcal{O}_Z(V)$ for an open set $V \subset Z$ as follows: a function $f : V \to C$ is in $\mathcal{O}_Z(V)$ if and only if there exists an open covering $V = \cup_i V_i$ in Z such that for each i we have $f_{|V_i} = g_{i|V_i}$ for some $g_i \in \mathcal{O}_X(U_i)$ where U_i is an open subset of X containing V_i. It is not difficult to check that $\mathcal{O}_Z$ is a sheaf of functions on Z. We will refer to it as the *induced structure sheaf* and denote it by $\mathcal{O}_{X|Z}$. Note that if Z is open in X then a subset $V \subset Z$ is open in Z if and only if it is open in X, and $\mathcal{O}_X(V) = \mathcal{O}_Z(V)$.

Let X be a topological space and let $X = \cup_i U_i$ be an open cover. Given sheaves of functions $\mathcal{O}_{U_i}$ on U_i for each i, which agree on each $U_i \cap U_j$, we can define a natural sheaf of functions $\mathcal{O}_X$ on X by "gluing" the $\mathcal{O}_{U_i}$. Let U be an open subset in X. Then $\mathcal{O}_X(U)$ consists of all functions on U, whose restriction to each $U \cap U_i$ belongs to $\mathcal{O}_{U_i}(U \cap U_i)$.

Let $(X, \mathcal{O}_X)$ be a geometric space. If $x \in X$ we denote by v_x the map from the set of C-valued functions on X to C obtained by evaluation at x:

$$v_x(f) = f(x).$$

Definition 1.2.5. A geometric space $(X, \mathcal{O}_X)$ is called an *abstract affine variety* if the following three conditions hold.

a) $\mathcal{O}_X(X)$ is a finitely generated C-algebra, and the map from X to the set $\mathrm{Hom}_C(\mathcal{O}_X(X), C)$ of C-algebra morphisms defined by $x \mapsto v_x$ is a bijection.

b) For each $f \in \mathcal{O}_X(X)$, $f \neq 0$, the set

$$X_f := \{x \in X : f(x) \neq 0\}$$

is open, and every nonempty open set in X is a union of some X_f's.

c) $\mathcal{O}_X(X_f) = \mathcal{O}_X(X)_f$, where $\mathcal{O}_X(X)_f$ denotes the C-algebra $\mathcal{O}_X(X)$ localized at f.

Remark 1.2.6. It can be checked that affine varieties with sheaves of regular functions are abstract affine varieties. We claim that, conversely, every abstract affine variety is isomorphic (as a geometric space) to an affine variety with its sheaf of regular functions. Indeed, let $(X, \mathcal{O}_X)$ be an abstract affine variety. Since $\mathcal{O}_X(X)$ is a finitely generated C-algebra, we can write $\mathcal{O}_X(X) = C[X_1, \dots, X_n]/I$ for some ideal I. As the elements in $\mathcal{O}_X(X)$ are C-valued functions on X, $\mathcal{O}_X(X)$ does not contain nonzero nilpotents; hence I is a radical ideal. By the Nullstellensatz (Theorem 1.1.6 c)), we can identify $\mathcal{O}_X(X)$ with the ring of regular functions $C[\mathcal{V}(I)]$ on $\mathcal{V}(I)$. Now a morphism of C-algebras from $C[\mathcal{V}(I)]$ to C corresponds to a maximal ideal of $C[\mathcal{V}(I)]$, hence to a point in $\mathcal{V}(I)$. Then, by the property a) of abstract affine varieties we can identify X with $\mathcal{V}(I)$ as a set. The Zariski topology on $\mathcal{V}(I)$ has the principal open sets as its base, so it now follows from b) that the identification of X and $\mathcal{V}(I)$ is a homeomorphism. Finally, by c), $\mathcal{O}_X(X_f)$ and the ring of regular functions on the principal open set X_f are also identified. This is enough to identify $\mathcal{O}_X(U)$ with the ring of regular functions on U for any open set U, as regularity is a local condition.

The preceding argument shows that the affine variety can be recovered completely from its algebra $\mathcal{O}_X(X)$ of regular functions, and conversely.

Example 1.2.7. In view of Remark 1.2.6, a closed subset of an abstract affine variety is an abstract affine variety (as usual, with the induced sheaf).

1.3. Projective varieties

In this section we define projective varieties as subsets of the projective space given by homogeneous polynomial equations. We shall see that a projective variety V has a finite open covering by affine varieties V_i and that the regular functions on V are determined by the regular functions on each V_i. This fact provides the model for the definition of algebraic varieties.

The *projective n-space* over C, denoted by $\mathbb{P}^n_C$, or $\mathbb{P}^n$, is the set of equivalence classes of $(n+1)$-tuples $(x_0, x_1, \dots, x_n)$ of elements in C not all zero under the equivalence relation $\sim$ defined by $(x_0, x_1, \dots, x_n) \sim (y_0, y_1, \dots, y_n) \Leftrightarrow y_i = \lambda x_i, 0 \leq i \leq n$, for some $\lambda \in C \setminus \{0\}$. Equivalently, $\mathbb{P}^n_C$ is the quotient set of $\mathbb{A}^{n+1}_C \setminus \{(0, \dots, 0)\}$ under the equivalence relation which identifies points lying on the same line through the origin. If V is a finite dimensional C-vector space, we define $\mathbb{P}(V)$ as the quotient of $V \setminus \{0\}$ by the equivalence relation $\sim$ defined by $v \sim w \Leftrightarrow v = \lambda w$ for some $\lambda \in C \setminus \{0\}$.

An element of $\mathbb{P}^n_C$ is called a point. We denote by $(x_0 : x_1 : \cdots : x_n)$ the equivalence class in $\mathbb{P}^n_C$ of the element $(x_0, x_1, \dots, x_n) \in \mathbb{A}^{n+1}_C \setminus \{(0, \dots, 0)\}$. We call $(x_0 : x_1 : \cdots : x_n)$ the homogeneous coordinates of the point P. They are defined up to a nonzero common factor in C.

Example 1.3.1. We can consider a map $\mathbb{A}^1 \to \mathbb{P}^1$ given by $x \mapsto (1 : x)$. It is injective and the image contains all points in $\mathbb{P}^1$ except $(0 : 1)$. This "extra point" is called point at infinity.

If we consider in the affine plane a line l and a point P not in l, then every line through P cuts l in a point, except the parallel line to l. Adding to l an extra point at infinity, we obtain a 1-to-1 correspondence between the set of lines through P and the set of points in l.

We will now define projective varieties in $\mathbb{P}^n_C$ in a way analogous to the definition of affine varieties. Due to the nonuniqueness of homogeneous coordinates, the fact that a polynomial in $C[X_0, X_1, \dots, X_n]$ vanishes at a point of $\mathbb{P}^n_C$ is not well determined for an arbitrary polynomial. We need to introduce homogeneous polynomials and homogeneous ideals of the ring $C[X_0, X_1, \dots, X_n]$. We shall define more generally the notion of a graded ring.

Definition 1.3.2. A *graded ring* is a commutative ring R together with a decomposition $R = \oplus_{d \geq 0} R_d$ of R into the direct sum of subgroups R_d such that for any $d, e \geq 0$, $R_d . R_e \subset R_{d+e}$. An element of R_d is called a *homogeneous element of degree d*. Thus any element of R can be written uniquely as a finite sum of homogeneous elements.

Example 1.3.3. The ring of polynomials $R = C[X_0, X_1, \dots, X_n]$ can be made into a graded ring by taking R_d to be the subgroup containing all linear combinations of monomials of total degree d in $X_0, X_1, \dots, X_n$. If $f \in R_d$, $\lambda \in C$, we have $f(\lambda x_0, \lambda x_1, \dots, \lambda x_n) = \lambda^d f(x_0, x_1, \dots, x_n)$; hence the fact that $(x_0 : x_1 : \dots : x_n)$ is a zero of f is well determined.

Definition 1.3.4. If $R = \oplus_{d \geq 0} R_d$ is a graded ring, an ideal $I \subset R$ is a *homogeneous ideal* if $I = \oplus_{d \geq 0} (I \cap R_d)$, i.e. if all homogeneous parts of every element of I also belong to I.

We define a *projective variety* as the set of common zeros in $\mathbb{P}^n_C$ of a finite collection of homogeneous polynomials in $C[X_0, X_1, \dots, X_n]$. To each homogeneous ideal I of $C[X_0, X_1, \dots, X_n]$ we associate the set $\mathcal{V}(I)$ of its common zeros in $\mathbb{P}^n_C$. Taking into account that $C[X_0, X_1, \dots, X_n]$ is a Noetherian ring and Exercise 13, each ideal of $C[X_0, X_1, \dots, X_n]$ has a finite set of homogeneous generators. Therefore the set $\mathcal{V}(I)$ is a projective variety.

Proposition 1.3.5. *The correspondence $\mathcal{V}$ satisfies the following equalities:*

a) $\mathbb{P}^n_C = \mathcal{V}(0), \emptyset = \mathcal{V}(C[X_0, X_1, \dots, X_n])$,

b) If I and J are two homogeneous ideals of $C[X_0, X_1, \dots, X_n]$, $\mathcal{V}(I) \cup \mathcal{V}(J) = \mathcal{V}(I \cap J)$,

c) If I_α is an arbitrary collection of homogeneous ideals of $C[X_0, X_1, \dots, X_n]$, $\cap_\alpha \mathcal{V}(I_\alpha) = \mathcal{V}(\sum_\alpha I_\alpha)$.

We define the *Zariski topology* in $\mathbb{P}^n_C$ as the topology having the projective varieties as closed sets.

Once we have a topological space, the notion of irreducible subset applies as in the affine case. The projective space $\mathbb{P}^n$ with the Zariski topology is a Noetherian space; hence we can define the dimension of a subset of $\mathbb{P}^n$ also as in the affine case.

Example 1.3.6. If f is a non constant homogeneous polynomial in the ring $C[X_0, X_1, \dots, X_n]$, $\mathcal{V}(f)$ is a *hypersurface in* $\mathbb{P}^n_C$. If f is a linear homogeneous polynomial, $\mathcal{V}(f)$ is a *hyperplane.*

Let us consider the map $\iota : \mathbb{A}^n \to \mathbb{P}^n$ given by

$$(x_1, \dots, x_n) \mapsto (1 : x_1 : \dots : x_n).$$

It is injective and its image is $\mathbb{P}^n \setminus H$, for $H := \{(x_0 : x_1 : \dots : x_n) : x_0 = 0\}$. The hyperplane H is called *hyperplane at infinity.*

Example 1.3.7. If $R = \oplus_{d \geq 0} R_d$ is a graded ring, $\oplus_{d>0} R_d$ is a homogeneous ideal of R. In particular, the ideal I_0 of $C[X_0, X_1, \dots, X_n]$ generated by $X_0, X_1, \dots, X_n$ is a proper radical homogeneous ideal of $C[X_0, X_1, \dots, X_n]$. We have $\mathcal{V}(I_0) = \emptyset$. This ideal I_0 is sometimes called the *irrelevant ideal* as it does not appear in the 1-1 correspondence between projective varieties in $\mathbb{P}^n_C$ and radical ideals of $C[X_0, X_1, \dots, X_n]$. (See Proposition 1.3.8 below.)

To each subset $S \subset \mathbb{P}^n_C$ we associate the homogeneous ideal $\mathcal{I}(S)$ generated by

$$\{f \in C[X_0, X_1, \dots, X_n] \ : \ f \text{ is homogeneous and } \ f(P) = 0 \text{ for all } \ P \in S\}.$$

As in the affine case, ideals of the form $\mathcal{I}(S)$ are radical ideals. If Y is a projective variety, we define the *homogeneous coordinate ring of* Y to be $C[Y] = C[X_0, X_1, \dots, X_n]/\mathcal{I}(Y)$.

We now state the projective Nullstellensatz, which has a similar formulation to the affine one, except for the adjustment due to the fact that the ideal I_0 defined in Example 1.3.7 has no common zeros. It can be easily deduced from the affine Nullstellensatz. The statement on irreducible varieties applies here as well.

Proposition 1.3.8. *The mappings $\mathcal{V}$ and $\mathcal{I}$ set a bijective correspondence between the closed subsets of $\mathbb{P}^n_C$ and the homogeneous radical ideals of $C[X_0, X_1, \dots, X_n]$ other than the irrelevant ideal I_0.*
Irreducible projective varieties correspond to homogeneous prime ideals under this correspondence. □

We shall now see that the projective space $\mathbb{P}^n$ has an open covering by affine n-spaces. Let H_i be the hyperplane $\{(x_0 : x_1 : \cdots : x_n) : x_i = 0\}$ and let U_i be the open set $\mathbb{P}^n \setminus H_i$. Then $\mathbb{P}^n$ is covered by the $U_i, 0 \leq i \leq n$ because if $P = (x_0 : x_1 : \cdots : x_n) \in \mathbb{P}^n$ at least one of the coordinates x_i is not zero, hence $P \in U_i$. We define a mapping

$$\begin{array}{cccc} \varphi_i : & U_i & \rightarrow & \mathbb{A}^n \\ & (x_0 : x_1 : \cdots : x_n) & \mapsto & \left(\frac{x_0}{x_i}, \ldots, \frac{x_{i-1}}{x_i}, \frac{x_{i+1}}{x_i}, \ldots, \frac{x_n}{x_i}\right). \end{array}$$

It is well defined as the ratios x_j/x_i are independent of the choice of homogeneous coordinates.

Proposition 1.3.9. *The map φ_i is a homeomorphism of U_i with its induced topology to $\mathbb{A}^n$ with its Zariski topology.*

Proof. The map φ_i is clearly bijective so it will be sufficient to prove that the closed sets of U_i are identified with the closed sets of $\mathbb{A}^n$ by φ_i. We may assume $i = 0$ and write U for U_0 and φ for φ_0. Let $R = C[X_0, X_1, \ldots, X_n], A = C[Y_1, \ldots, Y_n]$. We define a map α from the set R_h of homogeneous polynomials in R to A and a map β from A to R_h as follows. For $f \in R_h$, put $\alpha(f) = f(1, Y_1, \ldots, Y_n)$; for $g \in A$ of total degree e, the polynomial $X_0^e\, g(X_1/X_0, \ldots, X_n/X_0)$ is a homogeneous polynomial of degree e which we take as $\beta(g)$. From the definitions of α and β, we easily obtain the equalities $\varphi(\mathcal{V}(S)) = \mathcal{V}(\alpha(S))$, for a subset S of R_h, and $\varphi^{-1}(\mathcal{V}(T)) = \mathcal{V}(\beta(T))$, for a subset T of A. Hence both φ and φ^{-1} are closed maps, so φ is a homeomorphism. □

Corollary 1.3.10. *A subset S of $\mathbb{P}^n$ is closed if and only if its intersections $S \cap U_i$ are all closed (U_i being identified with $\mathbb{A}^n$ via the mapping φ_i defined above). If Y is a projective variety, then Y is covered by the open sets $Y \cap U_i, 0 \leq i \leq n$, which are homeomorphic to affine varieties via φ_i.* □

Remark 1.3.11. For a homogeneous polynomial f in $R = C[X_0, X_1, \ldots, X_n]$, the polynomial $\alpha(f) = f(1, Y_1, \ldots, Y_n) \in A = C[Y_1, \ldots, Y_n]$ is called the *dehomogenization of* f with respect to the variable X_0. For a polynomial $g \in A$ of total degree e, the polynomial $\beta(g) = X_0^e\, g(X_1/X_0, \ldots, X_n/X_0)$ is called the *homogenization of* g with respect to the new variable X_0. We can define analogously dehomogenization and homogenization with respect to the other variables. We can easily see $\alpha(\beta(g)) = g$ for any polynomial g in A. On the contrary, $\beta \circ \alpha$ is not the identity on R_d. For example, $\beta(\alpha(X_0)) = 1$. From the proof of Proposition 1.3.9, it follows that if Y is a projective variety, the ideal corresponding to $Y \cap U_0$, as an affine variety of U_0 identified with $\mathbb{A}^n$, is $\{\alpha(f) | f \in \mathcal{I}(Y)\}$.

Let V be a closed subset of $\mathbb{A}^n$. Identify $\mathbb{A}^n$ with one of the affine open subsets U_j of $\mathbb{P}^n$, say U_0. Then the closure $\overline{V}$ of V in $\mathbb{P}^n$ is called the *projective closure* of V. In particular, $\mathbb{P}^n$ is the projective closure of $\mathbb{A}^n$. The homogeneous ideal of $\overline{V}$ is then $\{\beta(g) | g \in \mathcal{I}(V)\}$. Since $V = U_0 \cap \overline{V}$, V is open in $\overline{V}$. Thus V is irreducible if and only if $\overline{V}$ is so.

Example 1.3.12. We consider the Fermat conic $V = \mathcal{V}(X^2+Y^2-Z^2) \subset \mathbb{P}^2_{\mathbb{C}}$. Its three affine pieces are

- $V_0 = \mathcal{V}(1 + Y^2 - Z^2)$, which is a hyperbola and has two points at infinity, namely $V \cap H_0 = \{(0:1:1), (0:1:-1)\}$.
- $V_1 = \mathcal{V}(X^2 + 1 - Z^2)$, which is again a hyperbola and has two points at infinity, namely $V \cap H_1 = \{(1:0:1), (1:0:-1)\}$.
- $V_2 = \mathcal{V}(X^2 + Y^2 - 1)$, which is a circle and has two points at infinity, namely $V \cap H_2 = \{(1:i:0), (1:-i:0)\}$, called the *circular points at infinity*, since they are the intersection of any circle in the affine plane with the line at infinity.

Example 1.3.13. Let $V = \mathcal{V}(Z - Y^2, T - Y^2)$ in $\mathbb{A}^3_{\mathbb{C}}$ identified with U_0. Its projective closure is $\overline{V} = V \cup \{(0:0:z:t) \in \mathbb{P}^3\}$. The homogeneous ideal of $\overline{V}$ is not generated by the homogenizations $XZ - Y^2$ and $XT - Y^2$ of the generators of the ideal of V as $Z - T$ is a homogeneous polynomial in $\mathcal{I}(V)$ which is not in $\langle XZ - Y^2, XT - Y^2 \rangle$.

Example 1.3.14. Let $V = \mathcal{V}(Y^2 - X^3 - a\,X - b)$ in $\mathbb{A}^2_{\mathbb{C}}$ identified with U_2. Its projective closure is $\overline{V} = V \cup \{(0:1:0)\}$; hence V has only one point at infinity.

We shall now see how to define functions on projective varieties. If $f, g \in C[X_0, X_1, \dots, X_n]$, then $F := f/g$ can be seen as a function on $\mathbb{P}^n$ (defined at the points where g does not vanish) only if f and g are homogeneous polynomials of the same degree, in which case we refer to F as a rational function of degree 0. If $g(P) \neq 0$ for some point $P \in \mathbb{P}^n$, we say that F is *regular* at P. Note that if a rational function of degree 0 is regular in some point, then it is regular in a neighborhood of this point. If U is a subset of a projective variety $V \subset \mathbb{P}^n$, a function $F : U \to C$ is *regular* if for any point P in U there exists an open neighborhood U' of P and homogeneous polynomials $f, g \in C[X_0, X_1, \dots, X_n]$ of the same degree such that $F = f/g$ in U' and $g(P) \neq 0$. If U is an open set, we write $\mathcal{O}_V(U)$ for the set of all regular functions on U.

Let U be an open subset of $\mathbb{P}^n$ contained in some of the affine open sets $U_i = \{(x_0 : x_1 : \dots : x_n) \in \mathbb{P}^n : x_i \neq 0\}$. Then U is also open in U_i, which is canonically identified with $\mathbb{A}^n$. We claim that $\mathcal{O}_{\mathbb{P}^n}(U) = \mathcal{O}_{\mathbb{A}^n}(U)$. We take, for example, $i = 0$. If $F \in \mathcal{O}_{\mathbb{P}^n}(U)$, for each $P \in U$

there exists an open neighborhood U' of P and homogeneous polynomials $f(X_0, X_1, \ldots, X_n), g(X_0, X_1, \ldots, X_n)$ of the same degree such that $F = f/g$ in U' and $g(P) \neq 0$. Then we also have in U',

$$F = f(1, Y_1, \ldots, Y_n)/g(1, Y_1, \ldots, Y_n),$$

where $Y_i = X_i/X_0, 1 \leq i \leq n$ are the affine coordinates on U_0. Hence $F \in \mathcal{O}_{\mathbb{A}^n}(U)$. Reciprocally, if $F \in \mathcal{O}_{\mathbb{A}^n}(U)$, on an open neighborhood U' of P we have $F = f/g$ and $g(P) \neq 0$, where $f, g \in C[Y_1, \ldots, Y_n]$. Then we also have in U',

$$F = \frac{X_0^{\max(\deg f, \deg g)} f(X_1/X_0, \ldots, X_n/X_0)}{X_0^{\max(\deg f, \deg g)} g(X_1/X_0, \ldots, X_n/X_0)},$$

which is the quotient of two homogeneous polynomials in $X_0, X_1, \ldots, X_n$ of the same degree; hence $F \in \mathcal{O}_{\mathbb{P}^n}(U)$. Analogously, if $V \subset \mathbb{P}^n$ is a projective variety, $V_i = V \cap U_i$, we have $\mathcal{O}_V(U) = \mathcal{O}_{V_i}(U)$ for every open subset U of V contained in V_i.

Exercises

(1) Prove the inclusion $\sqrt{I} \subset \mathcal{I}(\mathcal{V}(I))$ for an ideal I of $C[X_1, \dots, X_n]$.

(2) Provide the proof of Proposition 1.1.2.

(3) Provide the proof of Proposition 1.1.3.

(4) Prove that for a subset V of $\mathbb{A}^n_C$, $V = \mathcal{V}(\mathcal{I}(V))$ if and only if V is closed.

(5) Determine the open sets in the Zariski topology on $\mathbb{A}^1_C$. Prove that this topology is not Hausdorff.

(6) Determine the closed sets in $\mathbb{A}^2_C = \mathbb{A}^1_C \times \mathbb{A}^1_C$ with respect to the product topology (with $\mathbb{A}^1_C$ endowed with the Zariski topology) and with respect to the Zariski topology. In particular, obtain that $\Delta(\mathbb{A}^1_C) = \{(x, x) \in \mathbb{A}^1_C \times \mathbb{A}^1_C\}$ is a closed set in the latter topology but not in the former.

(7) Let $V \subset \mathbb{A}^n_C, W \subset \mathbb{A}^m_C$ be affine varieties. By identifying $\mathbb{A}^n_C \times \mathbb{A}^m_C$ with $\mathbb{A}^{n+m}_C$, we can consider the cartesian product $V \times W$ as a subset of $\mathbb{A}^{n+m}_C$. Prove that $V \times W$ is an affine variety and that there is an isomorphism

$$C[V \times W] \simeq C[V] \otimes C[W].$$

Hint: Assign to a pair $(g, h) \in C[V] \times C[W]$ the polynomial function $f(x, y) = g(x)h(y)$ on $V \times W$. This assignment induces a C-algebra morphism $C[V] \otimes C[W] \to C[V \times W]$.

(8) Let $V \subset \mathbb{A}^n_C, W \subset \mathbb{A}^m_C$ be closed irreducible sets. Prove that $V \times W$ is closed and irreducible in $\mathbb{A}^{n+m}_C$ with respect to the Zariski topology.
Hint: If $V \times W$ is the union of two closed subsets Z_1, Z_2, then $V = V_1 \cup V_2$, where $V_i := \{x \in V : \{x\} \times W \subset Z_i\}, i = 1, 2$, are closed in V.

(9) Provide the proof of the statements in Example 1.2.4.

(10) Check that an affine variety with its sheaf of regular functions is an abstract affine variety.

(11) Prove that $\mathbb{A}^2 \setminus \{(0, 0)\}$ is not an affine variety.

(12) Prove that every automorphism of $\mathbb{A}^1$ (=isomorphism from $\mathbb{A}^1$ into itself) has the form $x \mapsto ax + b \, (a \in C^*, b \in C)$.

(13) a) Prove that an ideal I of a graded ring $R = \oplus_{d \geq 0} R_d$ is homogeneous if and only if it is generated by homogeneous elements. Deduce that sums, products, intersections and radical of homogeneous ideals are again homogeneous.
b) Prove that if I is a homogeneous ideal of a graded ring R, then the quotient ring R/I is graded in a natural way.

(14) If I is an ideal of a graded ring $R = \oplus_{d\geq 0} R_d$, prove that

$$I_h := \oplus_{d\geq 0}(I \cap R_d)$$

is a homogeneous ideal of R. Prove that if I is prime, I_h is prime as well. Give an example of a prime ideal I of a graded ring such that $I \neq I_h$.

(15) Provide the proof of Proposition 1.3.5.

(16) For a homogeneous ideal $I \subset R = C[X_0, X_1, \ldots, X_n]$ show that the following conditions are equivalent.
a) $\mathcal{V}(I) = \emptyset$,
b) $\sqrt{I} =$ either R or the ideal $I_0 = \bigoplus_{d>0} R_d$,
c) $I \supset R_d$ for some $d > 0$.

(17) Prove the following statements.
a) If $S_1 \subset S_2$ are subsets of $\mathbb{P}^n$, then $\mathcal{I}(S_1) \supset \mathcal{I}(S_2)$.
b) For any two subsets S_1, S_2 of $\mathbb{P}^n$, $\mathcal{I}(S_1 \cup S_2) = \mathcal{I}(S_1) \cap \mathcal{I}(S_2)$.
c) For any subset S of $\mathbb{P}^n$, $\mathcal{V}(\mathcal{I}(S)) = \overline{S}$.

(18) Prove that if H is any hyperplane in $\mathbb{P}^n$, then $\mathbb{P}^n \setminus H$ is homeomorphic to $\mathbb{A}^n$.

(19) Provide the proof of Proposition 1.3.8.
Hint: Interpret the problem in terms of the affine space $\mathbb{A}_C^{n+1}$ and use the affine Nullstellensatz 1.1.6.

(20) If V is a projective variety with homogeneous coordinate ring $C[V]$, prove that $\dim C[V] = \dim V + 1$.

Chapter 2

Algebraic Varieties

In this chapter we define an algebraic variety as a geometric space with a finite open covering by affine varieties satisfying some separation axiom. We show that affine and projective varieties are algebraic varieties. We consider subvarieties and their dimensions and study morphisms of varieties.

2.1. Prevarieties

Definition 2.1.1. An *algebraic prevariety* is a geometric space $(X, \mathcal{O}_X)$ with the following property: X has a finite open covering $X = U_1 \cup \cdots \cup U_r$ such that each geometric space $(U_i, \mathcal{O}_{U_i})$ where $\mathcal{O}_{U_i}$ denotes the induced structure sheaf is an affine variety. For an open subset U of X, we call the elements in $\mathcal{O}_X(U)$ the *regular functions on* U.

Example 2.1.2. Each affine or projective variety with the corresponding sheaf of regular functions is a prevariety.

Lemma 2.1.3. *Let $(X, \mathcal{O}_X)$ be a prevariety with affine open covering $X = U_1 \cup \cdots \cup U_r$. We have*

a) X is a noetherian topological space.

b) Any open subset U of X is again a prevariety.

c) Any closed subset Z of X is again a prevariety.

Proof. a) follows from the fact that each U_i is noetherian.

b) We have $U = \cup_i (U \cap U_i)$ and $U \cap U_i$ is a open subset of the affine variety U_i, hence a finite union of principal open sets, which are affine.

c) We have $Z = \cup_i (Z \cap U_i)$ and the closed sets $Z \cap U_i$ of the affine varieties U_i are affine. □

A subset of a topological space X is called *locally closed* if it is the intersection of an open set with a closed set. It follows from the preceding lemma that a locally closed subset of a prevariety is again a prevariety. We shall refer to the locally closed subsets of a prevariety as *subprevarieties*.

A *morphism of prevarieties* will be a morphism with respect to their geometric space structure. If $\varphi : X \to Y$ is a morphism of prevarieties, V an open subset of Y, the assignment $f \mapsto f \circ \varphi$ is a C-algebra morphism $\mathcal{O}_Y(V) \to \mathcal{O}_X(\varphi^{-1}(V))$, which we denote by φ^*.

We now give a criterion to recognize when a map of prevarieties is a morphism.

Proposition 2.1.4. (Affine Criterion) *Let X, Y be prevarieties and $\varphi : X \to Y$ a map. Assume that there is an affine open covering $Y = \cup_{i=1}^r V_i$ and an open covering $X = \cup_{i=1}^r U_i$ such that*

a) $\varphi(U_i) \subset V_i$ for each i,

b) $f \circ \varphi \in \mathcal{O}_X(U_i)$ for each $f \in \mathcal{O}_Y(V_i)$.

Then φ is a morphism of geometric spaces.

Proof. An affine open covering of X induces one in each U_i. If U is an affine open subset of U_i, by b), we have that composing with φ sends $\mathcal{O}_Y(V_i) = C[V_i]$ into $\mathcal{O}_X(U) = C[U]$. So, by extending the index set if necessary, we reduce to the case where the U_i are also affine. Now by assumption $\varphi_i := \varphi_{|U_i} : U_i \to V_i$ is a morphism of affine varieties, since it is completely determined by the C-algebra morphism $\varphi_i^* : C[V_i] \to C[U_i]$. In particular φ_i is continuous, so φ is continuous.

Let $V \subset Y$ be an open set and $U := \varphi^{-1}(V)$. If $f \in \mathcal{O}_Y(V)$, b) implies that $f \circ \varphi \in \mathcal{O}_X(U \cup U_i)$, for $i = 1, \dots, r$. Since U is the union of the $U \cap U_i$ and since $\mathcal{O}_X$ is a sheaf (see Definition 1.2.1), we obtain $f \circ \varphi \in \mathcal{O}_X(U)$. □

We shall now define rational functions on an irreducible prevariety X. Consider pairs (U, f) where U is an open subset of X and $f \in \mathcal{O}_X(U)$. We call two such pairs (U, f) and (V, g) equivalent if $f = g$ on $U \cap V$. As X is irreducible, all nonempty open subsets of X are dense and then this relation is an equivalence relation. We denote by $\langle U, f \rangle$ the equivalence class of the pair (U, f) and by $C(X)$ the set of equivalence classes. As X is irreducible, any two nonempty open subsets of X intersect, so we can define in $C(X)$ addition and multiplication making $C(X)$ into a ring. Moreover, if $\langle U, f \rangle \in C(X)$ and $f \neq 0$, then we can restrict f to an open subset V of U where it does not vanish and obtain $\langle U, f \rangle^{-1} = \langle V, 1/f \rangle$. We call $C(X)$

the *function field* of X. It is easy to see that if X is affine this definition coincides with the one given before.

Let us consider a morphism of prevarieties $\varphi : X \to Y$. In case X, Y are irreducible and $\varphi(X)$ is dense in Y, the induced morphism $\varphi^* : \mathcal{O}_Y(V) \to \mathcal{O}_X(\varphi^{-1}(V))$, where V is an open subset of Y, can be thought of globally as a ring morphism $C(Y) \to C(X)$, whose restriction to $\mathcal{O}_Y(V)$ has image in $\mathcal{O}_X(\varphi^{-1}(V))$. Here φ^* is injective; hence we can see $C(X)$ as a field extension of $C(Y)$.

Let $\mathcal{F}$ be a sheaf of functions on a topological space X and $P \in X$. The open subsets of X containing P form an inverse system with respect to inclusion. The *stalk* $\mathcal{F}_P$ of $\mathcal{F}$ at P is defined to be the corresponding direct limit of the algebras $\mathcal{F}(U)$ via the restriction maps. The elements of the stalk $\mathcal{F}_P$ are called *germs of functions at* P. An element of $\mathcal{F}_P$ is represented by a pair (U, f), where U is an open neighborhood of P and $f \in \mathcal{F}(U)$. Two such pairs (U, f) and (V, g) represent the same element in $\mathcal{F}_P$ if there exists an open neighborhood W of P with $W \subset U \cap V$ such that $f_{|W} = g_{|W}$. If $(X, \mathcal{O}_X)$ is a prevariety, we write simply $\mathcal{O}_P$ for $(\mathcal{O}_X)_P$ and call it *the local ring at* P . It is indeed a local ring whose unique maximal ideal consists of the germs of functions vanishing at P.

If X is an irreducible affine variety, this definition agrees with the one given in Section 1.1.

We now look at the existence of products in the category of prevarieties. Given two prevarieties $(X, \mathcal{O}_X)$ and $(Y, \mathcal{O}_Y)$, we want to prove that there exists a prevariety $(Z, \mathcal{O}_Z)$ together with morphisms $\pi_1 : Z \to X, \pi_2 : Z \to Y$ such that the following universal property holds: if $(W, \mathcal{O}_W)$ is another prevariety with morphisms $\varphi_1 : W \to X, \varphi_2 : W \to Y$, then there exists a unique morphism $\psi : W \to Z$ such that $\pi_i \circ \psi = \varphi_i$, for $i = 1, 2$.

We first observe that, for prevarieties X and Y, the underlying set of a product of prevarieties is necessarily the cartesian product $X \times Y$. Indeed, by applying the universal property in the case in which W consists of a single point, we see that the points in Z correspond bijectively with pairs $(x, y) \in X \times Y$. To see how to give $X \times Y$ the structure of a prevariety, we first look at the affine case.

Proposition 2.1.5. *Let $X \subset \mathbb{A}^n, Y \subset \mathbb{A}^m$ be affine varieties. Endow the cartesian product $X \times Y$ with the Zariski topology (induced by the Zariski topology of $\mathbb{A}^{n+m} = \mathbb{A}^n \times \mathbb{A}^m$). Then*

a) $X \times Y$ with the projections $\pi_1 : X \times Y \to X$ and $\pi_2 : X \times Y \to Y$ is a categorical product of the prevarieties X, Y and $C[X \times Y] \simeq C[X] \otimes C[Y]$.

b) *If* $(x,y) \in X \times Y$, $\mathcal{O}_{(x,y)}$ *is the localization of* $\mathcal{O}_x \otimes \mathcal{O}_y$ *at the maximal ideal* $\mathfrak{M}_x \otimes \mathcal{O}_y + \mathcal{O}_x \otimes \mathfrak{M}_y$.

Proof. a) The statement on coordinate rings is proved in Exercise 7 of chapter 1. It remains to prove the universal property for $X \times Y$. Given a prevariety W and morphisms $\varphi_1 : W \to X, \varphi_2 : W \to Y$, we have to construct a suitable morphism $\psi : W \to X \times Y$. There is a unique such mapping of sets which makes $\varphi_i = \pi_i \circ \psi$. To check that it is a morphism, we use the Affine Criterion 2.1.4. As $X \times Y$ is affine, we just need to see that ψ pulls back polynomial functions on $X \times Y$ to regular functions on W. This follows from the fact that $C[X \times Y]$ is generated by the pullbacks of $C[X]$ and $C[Y]$ under the π_i and that the φ_i are morphisms.

b) If X, Y are irreducible, so is $X \times Y$ (Exercise 8, chapter 1). Part a) shows that $C[X] \otimes C[Y]$ is an integral domain, with fraction field isomorphic to $C(X \times Y)$. Now we have inclusions $C[X] \otimes C[Y] \subset \mathcal{O}_x \otimes \mathcal{O}_y \subset \mathcal{O}_{(x,y)}$. Since $\mathcal{O}_{(x,y)}$ is the localization of $C[X] \otimes C[Y]$ at the ideal $\mathfrak{M}_{(x,y)}$, it is also the localization of $\mathcal{O}_x \otimes \mathcal{O}_y$ at its ideal $\mathfrak{M}$ vanishing at (x,y). Evidently $\mathfrak{M}_x \otimes \mathcal{O}_y + \mathcal{O}_x \otimes \mathfrak{M}_y \subset \mathfrak{M}$. Conversely, let $f = \sum g_i \otimes h_i \in \mathfrak{M}$, with $g_i \in \mathcal{O}_x, h_i \in \mathcal{O}_y$. If $g_i(x) = a_i, h_i(y) = b_i$, then $f - \sum a_i b_i = \sum (g_i - a_i) \otimes h_i + \sum a_i \otimes (h_i - b_i) \in \mathfrak{M}_x \otimes \mathcal{O}_y + \mathcal{O}_x \otimes \mathfrak{M}_y$. Now $\sum a_i b_i = f(x,y) = 0$; hence $f \in \mathfrak{M}_x \otimes \mathcal{O}_y + \mathcal{O}_x \otimes \mathfrak{M}_y$. □

We now prove the existence of products in the case of arbitrary prevarieties.

Proposition 2.1.6. *Finite products exist in the category of prevarieties.*

Proof. To endow the cartesian product $X \times Y$ with the structure of prevariety, we have to specify a topology and a structure sheaf. For all affine open sets $U \subset X, V \subset Y$ and elements $h \in C[U] \otimes C[V]$, we decree that the principal open sets $(U \times V)_h$ should be basic open sets in $X \times Y$. Notice that these sets do form a basis for a topology as $(U_1 \times V_1)_{h_1} \cap (U_2 \times V_2)_{h_2} = ((U_1 \cap U_2) \times (V_1 \cap V_2))_{h_1 h_2}$. Moreover the description of the coordinate ring of $U \times V$ obtained in part a) of Proposition 2.1.5 gives that this topology coincides with the Zariski topology given in the affine case.

Now we define a structure sheaf on $X \times Y$. Let W be an open set in $X \times Y$ and f a C-valued map on W. Then we say that f is regular if and only if for each $x \in W$, there exists a basic open set $(U \times V)_h$ such that $f_{|(U \times V)_h} = a/h^m$, for some $a \in C[U] \otimes C[V]$ and some nonnegative integer m.

This defines a sheaf $\mathcal{O}_{X \times Y}$. Indeed, let $f \in \mathcal{O}_{X \times Y}(W)$ and let $W' \subset W$ be an open subset. For $x \in W'$, we have principal open sets $(U \times V)_h$ and

$(U' \times V')_{h'}$ such that $x \in (U \times V)_h \subset W$, $x \in (U' \times V')_{h'} \subset W'$ and $f_{|(U\times V)_h} = a/h^m$. Then $x \in (U \times V)_h \cap (U' \times V')_{h'} = ((U \cap U') \times (V \cap V'))_{hh'} \subset W'$ and

$$\frac{a}{h^m}|((U \cap U') \times (V \cap V'))_{hh'} = \frac{a'h'^m}{(hh')^m}$$

where a' denotes the restriction of a from $U \times V$ to $(U \cap U') \times (V \cap V')$, which belongs to $\mathcal{O}_X(U \cap U') \otimes \mathcal{O}_Y(V \cap V')$. So $f|W'$ is regular. As regularity is defined locally, the second axiom of sheaf is deduced easily.

Now we want to show that $(X \times Y, \mathcal{O}_{X\times Y})$ is a prevariety. We first check that the natural projections $\pi_1 : X \times Y \to X, \pi_2 : X \times Y \to Y$ are morphisms. They are continuous, as for an open subset U of X, we have $\pi_1^{-1}(U) = U \times Y$ which is open and, for $f \in \mathcal{O}_X(U)$, $\pi_1^*(f) = f \otimes 1 \in \mathcal{O}_X(U) \otimes \mathcal{O}_X(Y)$ is regular. Analogously for π_2. Now, if W is a prevariety, with morphisms $\varphi_1 : W \to X, \varphi_2 : W \to Y$, there is a unique map of sets $\psi : W \to X \times Y$ such that $\varphi_i = \pi_i \circ \psi$. We use the affine criterion 2.1.4 to prove that ψ is a morphism. By construction, products $U \times V$ of affine open sets U in X, V in Y, are affine open sets which cover $X \times Y$. Open sets of the form $\varphi_1^{-1}(U) \times \varphi_2^{-1}(V)$ cover W and the universal property of $U \times V$ shows that the restriction of ψ to such open sets is a morphism. □

2.2. Varieties

It is possible to find examples of prevarieties which are geometrically pathological. For example, let X be covered by two copies U, V of $\mathbb{A}^1$, with every point in the two copies identified except 0. So, X is the affine line with a point doubled. To avoid such cases, we make the following definition.

Definition 2.2.1. A prevariety X is called an *(algebraic) variety* if the diagonal $\Delta(X) = \{(x,x)|x \in X\}$ is closed in $X \times X$.

Remark 2.2.2. In the category of topological spaces, with $X \times X$ given the product topology, the condition $\Delta(X)$ closed in $X \times X$ is equivalent to the Hausdorff separation axiom.

Remark 2.2.3. An equivalent condition to $\Delta(X)$ closed in $X \times X$ is the following. For morphisms $\varphi, \psi : Y \to X$, where Y is any prevariety, the set $\{y \in Y : \varphi(y) = \psi(y)\}$ is closed in Y. Indeed, by applying this condition to the projections of $X \times X$ on both factors, we obtain $\Delta(X)$ closed in $X \times X$. In the other direction, the inverse image of $\Delta(X)$ by the morphism $Y \overset{\varphi\times\psi}{\to} X \times X$ is $\{y \in Y : \varphi(y) = \psi(y)\}$.

Let us see that the affine line with a point doubled is not a variety. If we take the two maps $\mathbb{A}^1 \to U \subset X, \mathbb{A}^1 \to V \subset X$ the subset of $\mathbb{A}^1$ on which they coincide is $\mathbb{A}^1 \setminus \{0\}$, which is not closed.

Example 2.2.4. **1.** An affine variety is a variety, as the diagonal is given by polynomials functions.

2. Subprevarieties of a variety are again varieties. These are therefore called *subvarieties.*

3. If X, Y are varieties, so is $X \times Y$.

Lemma 2.2.5. *Let X be a prevariety and assume that each pair $x, y \in X$ lie in some affine open subset of X. Then X is a variety.*

Proof. Given a prevariety Y and morphisms $\varphi, \psi : Y \to X$, let $Z = \{y \in Y : \varphi(y) = \psi(y)\}$. We have to show that Z is closed. If $z \in \overline{Z}$, $\varphi(z)$ and $\psi(z)$ lie by hypothesis in some affine open set V of X. Then $U = \varphi^{-1}(V) \cap \psi^{-1}(V)$ is an open neighborhood of z which must meet Z. But $Z \cap U = \{y \in U : \varphi'(y) = \psi'(y)\}$, where $\varphi', \psi' : U \to V$ are the restrictions. Since V is a variety, $Z \cap U$ is closed in U. This means that $U \setminus (Z \cap U)$ is an open set not meeting Z, so in particular it cannot contain z. We conclude that $z \in Z$. □

Corollary 2.2.6. *A projective variety is a variety.*

Definition 2.2.7. An open set in a projective variety with the induced sheaf of functions is called a *quasi-projective variety.*

The following proposition shows why it is better to deal with varieties than with prevarieties.

Proposition 2.2.8. *Let Y be a variety, X any prevariety.*

a) If $\varphi : X \to Y$ is a morphism, the graph $\Gamma_\varphi = \{(x, \varphi(x)) : x \in X\}$ is closed in $X \times Y$.

b) If $\varphi, \psi : X \to Y$ are morphisms which agree on a dense subset of X, then $\varphi = \psi$.

Proof. a) Γ_φ is the inverse image of $\Delta(Y)$ under the morphism $X \times Y \to Y \times Y$ which sends (x, y) to $(\varphi(x), y)$.

b) The set $\{x \in X : \varphi(x) = \psi(x)\}$ is closed in X since Y is a variety. It is dense by assumption, so it coincides with X. □

Definition 2.2.9. A morphism of affine varieties $\varphi : X \to Y$ is called *finite* if $C[X]$ is integral over the subring $\varphi^* C[Y]$.

Remark 2.2.10. If X, Y are irreducible and $\varphi : X \to Y$ is finite and dominant, then $C(X)$ is a finite algebraic extension of $\varphi^* C(Y)$ and so $\dim X = \dim Y$.

Proposition 2.2.11. *Let $\varphi : X \to Y$ be a finite, dominant morphism of affine varieties. If Z is closed in X, then $\varphi(Z)$ is closed in Y and the restriction of φ to Z is finite. In particular, φ is surjective.*

Proof. Let $R = C[X], S = C[Y]$. Since φ is dominant, φ^* is injective. We can then view S as a subring of R, over which R is integral by hypothesis. If I is an ideal of R, then R/I is an integral extension of $S/(I \cap S)$. Now let Z be closed in X, $I = \mathcal{I}(Z)$. Then φ maps Z into the zero set Z' of $I' = I \cap S$, which is a radical ideal of S, hence equal to $\mathcal{I}(Z')$. The corresponding coordinate rings are R/I and S/I', so by the remark above, $\varphi : Z \to Z'$ is again finite (and dominant). It now suffices to prove that any finite dominant morphism is surjective. If $y \in Y$, then to say that $\varphi(x) = y$ is just to say that φ^* sends the local ring of y into that of x, or that φ^* sends the maximal ideal M' of S vanishing at y into the maximal ideal M of R vanishing at x. To show that φ is surjective, we therefore have to show that M' lies in some maximal ideal M of R. But this follows from the Going Up theorem since R is integral over S.(See e.g. [**A-M**] 5.10.) □

We now extend the notion of dominant to morphisms of varieties.

Definition 2.2.12. A morphism of varieties $\varphi : V \to W$ is called *dominant* if $\varphi(\mathrm{dom}(\varphi))$ is a dense subset of W.

Proposition 2.2.13. *Let $\varphi : X \to Y$ be a morphism of varieties. Then $\varphi(X)$ contains a nonempty open subset of its closure $\overline{\varphi(X)}$.*

Proof. We first reduce to the case X, Y irreducible. Let $Y = \cup Y_i$ with Y_i the irreducible components of Y. Then $X_i := \varphi^{-1}(Y_i)$ are irreducible. The union of open sets of $\overline{\varphi(X_i)}$ contained in $\varphi(X_i)$ satisfies the statement in the proposition, so we may assume Y is irreducible. Now let $X = \cup X_i$ with X_i be the irreducible components of X. The union of open sets of $\overline{\varphi(X_i)}$ contained in $\varphi(X_i)$ satisfies the statement in the proposition, so we may assume X is irreducible. Let W be an affine open subset of Y meeting $\varphi(X)$ and consider $\varphi_{|\varphi^{-1}(W)}$. Then U satisfying the conditions for $\varphi_{|\varphi^{-1}(W)}$ satisfies them as well for φ. So we can assume that Y is affine. By substituting Y by $\overline{\varphi(X)}$, we can assume that φ is dominant. Then we must prove that $\varphi(X)$ contains a nonempty open subset of Y. We may also reduce to the case when X is affine. Indeed, let $X = \cup U_i$ be an open affine covering of X. As U_i is dense in X, $\varphi(U_i)$ is dense in Y, so the restrictions $\varphi_{|U_i} : U_i \to Y$ are dominant morphisms of irreducible affine varieties. Now the union of open subsets of $\overline{\varphi(U_i)}$ contained in $\varphi(U_i)$ will satisfy the statement.

Let $R = C[X], S = C[Y]$. As φ^* is injective, we can consider S as a subring of R. Now by Corollary 1.1.10, we can find elements $f \in S$, $x_1, \dots, x_r \in R$, algebraically independent over S_f and such that R_f is integral over $S_f[x_1, \dots, x_r]$. Now $R_f = C[X_f], S_f = C[Y_f]$ where X_f, Y_f denote the principal open sets in X and Y defined by the nonvanishing of f. So $S_f[x_1, \dots, x_r] \simeq S_f \otimes C[x_1, \dots, x_r]$ can be viewed as the coordinate ring of $Y_f \times \mathbb{A}^r$. The restriction of φ to X_f can be factored as $X_f \overset{\psi}{\to} Y_f \times \mathbb{A}^r \overset{pr_1}{\to} Y_f$, where ψ is a finite dominant morphism. The principal open set Y_f satisfies $\varphi^{-1}(Y_f) = X_f$.

By Proposition 2.2.11, ψ is surjective, as is pr_1. Therefore Y_f lies in $\varphi(X)$. □

A variety X is a noetherian topological space, so, as in the affine case, we define the dimension of a variety to be its dimension as a topological space. If X is irreducible, its field $C(X)$ of rational functions coincides with $C(U)$ for any affine open subset U of X. As U is dense in X, we have $\dim X = \dim U$. Hence, by the affine case, we obtain that the dimension of X is equal to the transcendence degree of $C(X)$ over C.

Similarly, if X is an irreducible variety, $f \in C(X)$, we have $\dim X_f = \dim X$. For example, by identifying the set of square matrices of order n with entries in C with the affine space $\mathbb{A}^{n^2}$, we can consider the subset $\mathrm{GL}(n, C)$ of matrices with nonzero determinant as a principal open subset of $\mathbb{A}^{n^2}$. We then have $\dim \mathrm{GL}(n, C) = n^2$.

We now look at the dimension of subvarieties.

Proposition 2.2.14. *Let X be an irreducible variety, Y a proper, closed, irreducible subset. Then* $\dim Y < \dim X$.

Proof. We can assume that X is affine and let $d = \dim X$. Then $C[Y] \simeq C[X]/P$ for P a nonzero prime ideal of $C[X]$. By Noether's normalization lemma, we can choose transcendence bases of $C(X)$ and $C(Y)$ contained in $C[X]$ and $C[Y]$ respectively. Suppose $\dim Y \geq d$ and let $y_1, \dots, y_d$ algebraically independent elements in $C[Y]$. Then their preimages $x_1, \dots, x_d$ in $C[X]$ are clearly algebraically independent as well. Let f be a nonzero element in P. As $\dim X = d$, there must be a nontrivial polynomial relation $g(f, x_1, \dots, x_d) = 0$, where $g \in C[T_0, T_1, \dots, T_d]$. As $f \neq 0$, we can assume that $h(T_1, \dots, T_d) = g(0, T_1, \dots, T_d)$ is a nonzero polynomial. Now $h(y_1, \dots, y_d) = 0$, contradicting the independence of the y_i. □

Definition 2.2.15. We define the *codimension* $\operatorname{codim}_X Y$ of a subvariety Y of the variety X as $\operatorname{codim}_X Y = \dim X - \dim Y$.

We shall prove that the irreducible subvarieties of codimension 1 are precisely the irreducible components of hypersurfaces.

Corollary 2.2.16. *Let X be an irreducible affine variety, Y a closed irreducible subset of codimension 1. Then Y is a component of $\mathcal{V}(f)$ for some $f \in C[X]$.*

Proof. By assumption $Y \neq X$, so there exists a non zero $f \in C[X]$ vanishing on Y. Then $Y \subset \mathcal{V}(f) \subsetneq X$. Let Z be an irreducible component of $\mathcal{V}(f)$ containing Y. Proposition 2.2.14 says that $\dim Z < \dim X$, while $\dim Y \leq \dim Z$ with equality only if $Y = Z$. Since $\operatorname{codim}_X Y = 1$, equality must hold. □

In the situation of the corollary, it is not usually possible to make Y be precisely $\mathcal{V}(f)$. However this can be done when Y has codimension 1 in some affine space $\mathbb{A}^n$ or more generally when $C[X]$ is a unique factorization domain. (See Exercise 6.)

We shall now see a converse to Corollary 2.2.16.

We defined a hypersurface in $\mathbb{A}^n$ as the zero set of a single nonconstant polynomial $f \in C[X_1, \ldots, X_n]$. More generally, when V is an affine variety, a nonzero nonunit $f \in C[V]$ defines a hypersurface in V. For example $\mathrm{SL}(n, C)$ is the hypersurface in $\mathrm{GL}(n, C)$ or in $\mathbb{A}^{n^2}$ defined by $\det(X_{ij}) = 1$.

Proposition 2.2.17. *All irreducible components of a hypersurface in $\mathbb{A}^n$ have codimension 1.*

Proof. It suffices to consider the zero set V of an irreducible polynomial $f \in C[X_1, \ldots, X_n]$. As f is nonconstant, at least one variable, say X_n, actually occurs in f. Let x_i be the restriction of X_i to V, so $C(V) = C(x_1, \ldots, x_{n-1})$. We claim that $x_1, \ldots, x_{n-1}$ are algebraically independent over C. Otherwise there exists a nontrivial polynomial relation $g(x_1, \ldots, x_{n-1}) = 0$, whence $g(X_1, \ldots, X_{n-1})$ vanishes on V. As $\mathcal{I}(V) = (f)$, g would be a multiple of f, which is impossible as X_n occurs in f. We conclude that $\dim V \geq n - 1$, which must be an equality by Proposition 2.2.14. □

We now generalize this result to arbitrary affine varieties.

Theorem 2.2.18 (Krull's Hauptidealsatz)**.** *Let V be an irreducible affine variety, f a nonzero nonunit in $C[V]$, Y an irreducible component of $\mathcal{V}(f)$. Then Y has codimension 1 in V.*

Proof. Let $\mathfrak{p} = \mathcal{I}(Y) \subset R = C[V]$ and let $Y_1, \dots, Y_t$ be the components of $\mathcal{V}(f)$ other than Y, $\mathfrak{p}_i = \mathcal{I}(Y_i)$. The Nullstellensatz 1.1.6 implies that $\sqrt{Rf} = \mathfrak{p} \cap \mathfrak{p}_1 \cap \cdots \cap \mathfrak{p}_t$. Choose $g \in \mathfrak{p}_1 \cap \cdots \cap \mathfrak{p}_t \setminus \mathfrak{p}$ (such a g exists as Y is not contained in $Y_1 \cup \cdots \cup Y_t$). Then V_g is an irreducible affine variety having the same dimension as V and $Y \cap V_g$ is precisely the zero set of f in V_g. Since $Y \cap V_g$ is a principal open set in Y, it suffices to prove that its codimension in V_g is 1. So we may assume $Y = \mathcal{V}(f), \mathfrak{p} = \sqrt{Rf}$.

Now we apply Noether's normalization Lemma 1.1.8 to the domain $R = C[V]$. We obtain that R is integral over a subring S which is isomorphic to the polynomial ring $C[T_1, \dots, T_d]$, where $d = \dim V$. Let $E = C(V)$ and F the field of fractions of S. Since the field extension $E|F$ is finite, we may consider its norm $N_{E|F}$. As R is integral over S, $N_{E|F}$ takes elements of R into S.

Set $f_0 = N_{E|F}(f)$. We want to see $\sqrt{f_0 S} = \sqrt{fR} \cap S$. As f is integral over S, we have a relation $f^k + a_1 f^{k-1} + \cdots + a_k = 0$. By the properties of the norm, we have $f_0 = a_k^m$ (where $m = [E : F(f)]$). Then $0 = (f^k + a_1 f^{k-1} + \cdots + a_k)\, a_k^{m-1} = f\, a_k^{m-1}(f^{k-1} + a_1 f^{k-2} + \cdots + a_{k-1}) + f_0$, so $f_0 \in fR$. This implies $\sqrt{f_0 S} \subset \sqrt{fR} \cap S$. Now let $g \in \sqrt{fR} \cap S$, so $g^r = f\,h$, for some positive integer r and some $h \in R$. Taking norms, we obtain $g^{r[E:F]} = N_{E|F}(f) N_{E|F}(h)$. As $N_{E|F}(h) \in S$, we obtain $g \in S f_0$.

We have replaced the prime ideal $\mathfrak{p} = \sqrt{fR}$ by the prime ideal $\mathfrak{p} \cap S = \sqrt{f_0 S}$. Now S is a unique factorization domain and it is easy to see that if two different irreducible factors appear in the decomposition of f_0, then $\sqrt{f_0 S}$ would not be prime. We obtain then that, up to a unit factor, f_0 is a power of a nonconstant irreducible polynomial p; whence $\mathfrak{p} \cap S = pS$.

If S is viewed as the affine algebra of $\mathbb{A}^d$, then $\mathfrak{p} \cap S$ defines a hypersurface in $\mathbb{A}^d$, which has codimension 1, by Proposition 2.2.17. This means that the fraction field of $S/(\mathfrak{p} \cap S)$ has transcendence degree $d-1$ over C. On the other hand, R integral over S clearly implies $R/\mathfrak{p}$ integral over $S/(\mathfrak{p} \cap S)$, so the two fraction fields have equal transcendence degree. As the fraction field of $R/\mathfrak{p}$ is $C(Y)$, we obtain $\dim Y = d - 1$. □

Corollary 2.2.19. *Let V be an irreducible variety, U an open subset in V, f a nonzero nonunit in $\mathcal{O}_V(U)$, and Y an irreducible component of $\mathcal{V}(f)$. Then Y has codimension 1 in V.*

Proof. Just cut down with an affine open subset of U meeting Y and apply Theorem 2.2.18. □

Definition 2.2.20. Let us recall that a subset of a topological space X is called locally closed if it is the intersection of an open set with a closed set. A finite union of locally closed sets is called a *constructible set*.

Theorem 2.2.21 (Chevalley theorem). *Let $\varphi : X \to Y$ be a morphism of varieties. Then φ maps constructible sets to constructible sets. In particular, $\varphi(X)$ is constructible in Y.*

Proof. A locally closed subset of a variety is itself a variety, so it suffices to prove that $\varphi(X)$ is constructible. We may assume that Y is irreducible. We proceed by induction on $\dim Y$. For $\dim Y = 0$, there is nothing to prove. By induction, we can assume φ dominant.

Choose an open subset U of Y contained in $\varphi(X)$, using Proposition 2.2.13. Then the irreducible components $W_1, \dots, W_t$ of $Y \setminus U$ have smaller dimension than Y (Proposition 2.2.14). By induction, the restrictions of φ to $Z_i = \varphi^{-1}(W_i)$, $1 \leq i \leq t$, have images which are constructible in W_i, hence also constructible in Y. Therefore $\varphi(X)$ is constructible, being the union of U and the finitely many $\varphi(Z_i)$. □

The next proposition will be used in the construction of the quotient of an algebraic group by a subgroup. To prove it we need two lemmas.

Lemma 2.2.22. *Let X, Y be affine varieties such that $C[X] = C[Y][f]$, for some element $f \in C[X]$. We consider the morphism $\varphi : X \to Y$ defined by the inclusion $C[Y] \hookrightarrow C[X]$. Assume that f is transcendental over $C(Y)$. Then*

a) φ is an open morphism.

b) If Y' is an irreducible closed subvariety of Y, then $\varphi^{-1}(Y')$ is an irreducible closed subvariety of X of dimension equal to $\dim Y' + 1$.

Proof. a) We may assume that $X = Y \times \mathbb{A}^1$ and that φ is the projection on the first factor. Let $g = \sum_{i=0}^r g_i f^i \in C[X] = C[Y][f]$. Then $\varphi(X_g) = \cup_{i=0}^r Y_{g_i}$. As the image of principal open sets is open, φ is open.
b) Let $Q = \mathcal{I}(Y') \subset C[Y]$. Then $\varphi^{-1}(Y') = \mathcal{V}(P)$, for $P = QC[X]$. Then $C[X]/P \simeq (C[Y]/Q)[f]$. Since the last ring is an integral domain, P is a prime ideal and $\varphi^{-1}(Y')$ is irreducible. As $C[\varphi^{-1}(Y')] \simeq C[Y'][f]$, with f transcendental over $C(Y')$, $\dim \varphi^{-1}(Y') = \dim Y' + 1$. □

Lemma 2.2.23. *Let X, Y be affine varieties such that $C[X] = C[Y][f]$, for some element $f \in C[X]$. We consider the morphism $\varphi : X \to Y$ defined by the inclusion $C[Y] \hookrightarrow C[X]$. Assume that f is algebraic over $C(Y)$. Then there is a nonempty open subset U of X with the following properties.*

a) The restriction of φ to U defines an open morphism $U \to Y$.

b) If Y' is an irreducible closed subvariety of Y and X' is an irreducible component of $\varphi^{-1}Y'$ that intersects U, then $\dim X' = \dim Y'$.

c) *For* $x \in U$ *the fiber* $\varphi^{-1}(\varphi(x))$ *is a finite set with* $[C(X) : C(Y)]$ *elements.*

Proof. We have $C[X] = C[Y][T]/I$, where I is the ideal of $C[Y][T]$ of polynomials in the variable T vanishing at f. Let F be the minimal polynomial of f over $C(Y)$. Let a be a common denominator of the coefficients of F, so $F \in C[Y]_a[T]$. Let $f_1, \ldots, f_n$ be the roots of F in some extension field of $C(Y)$. Since $\operatorname{char} C(Y) = 0$, the roots are distinct and the discriminant $d = \prod_{i<j}(f_i - f_j)^2$ is a nonzero element of $C(Y)$, which can be expressed polynomially in the coefficients of F. It follows that there exist $b \in C[Y]$ and $m \geq 0$ such that $a^m d = b$.

We may replace X and Y by the principal open sets X_{ab} and Y_{ab}, respectively. We are then reduced to prove the lemma when, moreover, the following hold.

1. I contains the minimal polynomial F of f. From this it follows, using the division algorithm that I is the ideal generated by F. It also follows that $C[X]$ is a free $C[Y]$-module.
2. If $F(T) = \sum_{i=0}^n h_i T^i$, then for all $y \in Y$ the polynomial $F(y)(T) = \sum_{i=0}^n h_i(y) T^i$ has distinct roots.

We shall show that in this situation the statements of the lemma hold with $U = X$. We may assume that

$$X = \{(y,t) \in Y \times \mathbb{A}^1 \,:\, F(y)(t) = 0\},$$

and that the morphism $\varphi : X \to Y$ is the first projection. Let $G \in C[Y][T]$ and denote by g its class in $C[X]$. Then

$$X_g = \{(y,t) \in X \,:\, G(y)(t) \neq 0\}.$$

Write $G = QF + R$, where $R = \sum_{i=0}^{n-1} r_i T^i$ is a polynomial in T of degree $< n = \deg F$. Then $\varphi(X_g)$ is the set of $y \in Y$ such that not all roots of $F(y)(T)$ are roots of $R(y)(T)$. Since the first polynomial has n distinct roots, this is the set of $y \in Y$ such that $R(y)$ is not the zero polynomial. We then have

$$\varphi(X_g) = \bigcup_{i=0}^{n-1} Y_{r_i},$$

whence a).

Next let Y' as in b) and let $Q = \mathcal{I}(Y')$. Then $\varphi^{-1}(Y') = \mathcal{I}(QC[X])$. Let $A = C[Y]/Q$ and denote by $\overline{F}$ the image of F in $A[T]$. We claim that $QC[X]$ is a radical ideal, i.e. $A[T]/(\overline{F})$ has no nonzero nilpotent elements. Let $H \in A[T]$ and assume that H^m is divisible by $\overline{F}$. We may assume that

$\deg H < n$. It follows from Property 2 that $\overline{F}$ has distinct roots and that H is divisible by $\overline{F}$ as polynomials with coefficients in the quotient field of A. But, since H has lower degree than $\overline{F}$, this can only be if $H = 0$, which implies the claim. As $QC[X]$ is radical, it is an intersection of prime ideals of $C[X]$, say $QC[X] = \bigcap_{i=1}^{r} P_i$. We may assume that there are no inclusions among the P_i. The irreducible components of $\varphi^{-1}(Y')$ are the $\mathcal{V}(P_i)$. We show that $P_i \cap C[Y] = Q, 1 \leq i \leq r$. If this is not so, we have, say $P_1 \cap C[Y] \neq Q$. Take $x_1 \in (P_1 \cap C[Y]) \setminus Q$ and $x_i \in P_i \setminus P_1$ $(2 \leq i \leq r)$. Then $x_1 x_2 \dots x_r \in QC[X]$. Since $C[X]$ is free over $C[Y]$, it follows that $x_2 \dots x_r \in QC[X] \setminus P_1$, which is impossible if $r > 1$. If $r = 1$, we have a contradiction, since $QC[X] \cap C[Y] = Q$.

It follows that the quotient field of $C[X]/P_i$ is an algebraic extension of the quotient field of A, which proves b).

If Y' is a point then Q is a maximal ideal of $C[Y]$ and $A = C$. The preceding analysis shows that now $\varphi^{-1}(Y')$ is the zero dimensional variety defined by the C-algebra $C[T]/(\overline{F})$. Since $\overline{F}$ is a polynomial of degree n with distinct roots, c) follows. □

Proposition 2.2.24. *Let X and Y be irreducible varieties and let $\varphi : X \to Y$ be a dominant morphism. Let $r := \dim X - \dim Y$. There is a nonempty open subset U of X with the following properties.*

a) The restriction of φ to U is an open morphism $U \to Y$.

b) If Y' is an irreducible closed subvariety of Y and X' an irreducible component of $\varphi^{-1}(Y')$ that intersects U, then $\dim X' = \dim Y' + r$. In particular, if $y \in Y$, any irreducible component of $\varphi^{-1}y$ that intersects U has dimension r.

c) If $C(X)$ is algebraic over $C(Y)$, then for all $x \in U$ the number of points of the fiber $\varphi^{-1}(\varphi x)$ equals $[C(X) : C(Y)]$.

Proof. Assume that we have a factorization $\varphi = \varphi_1 \circ \varphi_2$, where $\varphi_2 : X \to Z, \varphi_1 : Z \to Y$ are dominant morphisms and Z is irreducible. If a) and b) hold for φ_1 and φ_2, they also hold for φ. To prove the theorem, we can assume that X and Y are affine. Since $C[X]$ is a $C[Y]$-algebra of finite type, we can find a factorization of φ

$$X = X_r \xrightarrow{\varphi_r} X_{r-1} \xrightarrow{\varphi_{r-1}} \cdots \xrightarrow{\varphi_2} X_1 \xrightarrow{\varphi_1} X_0 = Y,$$

where each φ_i is a morphism of the type of the morphism φ considered either in Lemma 2.2.22 or in Lemma 2.2.23. In particular, when $C(X)$ is algebraic over $C(Y)$, every φ_i is of the type considered in Lemma 2.2.23. We then obtain the proposition from these lemmas. □

Corollary 2.2.25. *In Proposition 2.2.24, a) can be replaced by the following stronger property:*

a') For any variety Z, the restriction of φ to U defines an open morphism $U \times Z \to Y \times Z$.

Proof. It suffices to prove this for Z affine. Observe that if a') holds for Z and Z' is a closed subvariety of Z, then a') also holds for Z'. Hence it suffices to establish a') for $Z = \mathbb{A}^m$. This will follow if we prove the corresponding result in the cases of Lemma 2.2.22 and Lemma 2.2.23. The first case is trivial. For the second one, a') follows by observing that if F is the minimal polynomial over $C(Y)$ of an element f in $C(X)$, then F is also the minimal polynomial of f over $C(Y \times \mathbb{A}^m)$. □

In 2.2.9 we defined finite morphism of affine varieties. We now see that a finite morphism has finite fibers. Let us recall that if B is an A-algebra of finite type, then B is finite over A if and only if B is integral over A.

Proposition 2.2.26. *Let $\varphi : X \to Y$ be a finite morphism of affine varieties.*

a) φ is closed.

b) $\varphi^{-1}(y)$ is finite for all $y \in Y$.

Proof. Let $A = C[X], B = C[Y]$.

a) Let $Z = \mathcal{V}(I)$ be a closed subset of X and $J = \varphi^{*-1}(I)$. The points of $\varphi(Z)$ are $\varphi^{*-1}(\mathfrak{M})$, with $\mathfrak{M}$ a maximal ideal of A containing I. Identifying B/J with a subring of A/I via φ^*, A/I is a finite B/J-algebra. For $\mathfrak{N}$ a maximal ideal of B containing J, $\mathfrak{N}' = \mathfrak{N}/J$ is a maximal ideal of B/J and by the Going Up theorem, there exists a maximal ideal $\mathfrak{M}'$ of A/I such that $\mathfrak{N}' = \mathfrak{M}' \cap (B/J)$. Let $\mathfrak{M}$ be the maximal ideal of A containing I corresponding to $\mathfrak{M}'$. We have $\mathfrak{N} = \varphi^{*-1}(\mathfrak{M})$. It follows that $\varphi(Z) = \mathcal{V}(J)$; hence $\varphi(Z)$ is closed in Y.

b) By a), $\varphi(X) = \mathcal{V}(\operatorname{Ker} \varphi^*)$. As A is a finite $(B/\operatorname{Ker} \varphi*)$-algebra, there is a finite number of maximal ideals of A lying above a maximal ideal of $B/(\operatorname{Ker} \varphi^*)$. So the result follows. □

Definition 2.2.27. Let $\varphi : X \to Y$ be a morphism of affine varieties. We say that φ is *locally finite* at a point $x \in X$ if there exist a finite morphism $\mu : Y' \to Y$ and an isomorphism ν of an open neighborhood U of x onto an open set in Y' such that $\mu \circ \nu$ is the restriction of φ to U.

Lemma 2.2.28. *Let $\varphi : X \to Y, \psi : Y \to Z$ be morphisms of affine varieties. If φ is locally finite in x and ψ is so in $\varphi(x)$, then $\psi \circ \varphi$ is locally finite in x.*

Proof. We may assume that Y is the principal open set Z'_f, where Z' is finite over Z, with $f \in C[Z']$. If Y' is finite over Y, then $C[Y'] = B_f$, where B is integral over $C[Z']$. Hence B is integral over $C[Z]$. It follows that Y' is isomorphic to a principal open set V_g, where V is finite over Z, with $g \in C[V]$. □

Lemma 2.2.29. *Let $\varphi : X \to Y$ be a dominant morphism of irreducible varieties. We consider $B = C[Y]$ as a subring of $A = C[X]$. Assume that $A = B[a]$, for some $a \in A$. Let $x \in X$. Then one of the two following statements holds:*

a) $\varphi^{-1}(\varphi x)$ is finite and φ is locally finite in x,

b) $\varphi^{-1}(\varphi x) \simeq \mathbb{A}^1$.

Proof. We have $A = B[T]/I$, where I is the ideal of the polynomials $f \in B[T]$ with $f(a) = 0$. Let $\varepsilon : B \to C$ be the morphism defining φx. It extends to a morphism $B[T] \to C[T]$. If $\varepsilon(I) = 0$, then $C[\varphi^{-1}(\varphi x)] \simeq C[T]$; whence $\varphi^{-1}(\varphi x) \simeq \mathbb{A}^1$.

If $\varepsilon(I) \neq 0$, the polynomials in $\varepsilon(I)$ vanish in $a(x)$; hence $\varepsilon(I)$ contains non-constant polynomials and no non-zero constants. This implies that $\varphi^{-1}(\varphi x)$ is finite. It also follows that there is $f \in I$ of the form $f_n T^n + \cdots + f_m T^m + \cdots + f_0$, where $\varepsilon(f_n) = \cdots = \varepsilon(f_{m+1}) = 0, \varepsilon(f_m) \neq 0, m > 0$. Put $s = f_n a^{n-m} + \cdots + f_m$. Then $s \neq 0$ and $s\, a^m + f_{m-1} a^{m-1} + \cdots + f_0 = 0$. We have then that sa is integral over $B[s]$ and a is integral over the subring $B[s^{-1}]$ of the quotient field of B. But since $s \in B[a]$, it follows that s is integral over $B[s^{-1}]$, i.e. that s is integral over B. Now the assertion of a) follows by observing that $A_s = B[sa, s]_s$. □

Proposition 2.2.30. *Let $\varphi : X \to Y$ be a dominant morphism of irreducible varieties. We consider $B = C[Y]$ as a subring of $A = C[X]$. Let $x \in X$. If the fiber $\varphi^{-1}(\varphi x)$ is finite, then φ is locally finite in x. Moreover* $\dim X = \dim Y$.

Proof. We have $A = B[a_1, \dots, a_h]$. If $h = 1$, the assertion is true by Lemma 2.2.29. We have a factorization of φ: $X \overset{\psi}{\to} X' \overset{\varphi'}{\to} Y$, where $C[X'] = B[a_1]$. Clearly $\psi^{-1}(\psi x)$ is finite. By induction on h we may assume that ψ is locally finite in x. We may then assume that there is a finite morphism of affine varieties $\psi' : X'' \to X'$ such that X is an affine open subset of X'' and that ψ is induced by ψ'.

Let $F = \varphi'^{-1}(\varphi' x)$. If F is infinite, by 2.2.29, it is isomorphic to $\mathbb{A}^1$. Let E be a component of $\psi'^{-1}(F)$ of dimension ≥ 1 passing through x. Now $X \cap E$ is an open subset of E containing x, hence must be infinite. But $X \cap E$ lies in the finite set $\varphi^{-1}(\varphi x)$ and we get a contradiction. Hence the components of $\psi'^{-1}(F)$ of dimension ≥ 1 do not contain x. Replacing X by a suitable open neighborhood of x we may assume that no such component exists. Then F is finite. The theorem then follows by using Lemmas 2.2.28 and 2.2.29. □

Definition 2.2.31. A point x of an irreducible variety X is *normal* if there exists an affine open neighborhood U of x such that $C[U]$ is integrally closed. A variety is normal if all its points are normal.

We now state a version of Zariski's main theorem.

Theorem 2.2.32. *Let $\varphi : X \to Y$ be a morphism of irreducible varieties that is bijective and birational. Assume Y to be normal. Then φ is an isomorphism.*

Proof. Let $x \in X$. Replace X and Y by affine open neighborhoods U of x, respectively V of φx. We deduce from Proposition 2.2.30 that we may assume that U is isomorphic to an affine open subset of an affine variety V' which is finite over V. By the birationality assumption, $C(V') \simeq C(V)$. Now the normality of Y implies that the finite morphism $V' \to V$ is in fact an isomorphism. This shows that φ is an isomorphism of geometric spaces, hence an isomorphism of varieties. □

We shall now define the tangent space of a variety V at a point x. If $f(X_1, \ldots, X_n) \in C[X_1, \ldots, X_n]$, $x = (x_1, \ldots, x_n)$ is a point in $\mathbb{A}^n$, we define the *differential* of f at x as

$$d_x f = \sum_{i=1}^{n} (\partial f / \partial X_i)(x)(X_i - x_i).$$

It follows from the definition that for $f, g \in C[X_1, \ldots, X_n]$, $d_x(f + g) = d_x f + d_x g$ and $d_x(fg) = f(x) d_x g + g(x) d_x f$.

If V is an affine variety in $\mathbb{A}^n_C$, x a point in V, we define the *tangent space* to V at the point x as the linear variety $\operatorname{Tan}(V)_x \subset \mathbb{A}^n_C$ defined by the vanishing of all $d_x f$, for $f \in \mathcal{I}(V)$. It is easy to see that for any finite set of generators of $\mathcal{I}(V)$, the corresponding $d_x f$ generate $\mathcal{I}(\operatorname{Tan}(V)_x)$. Notice that the tangent space to a linear variety is the variety itself at any of its points.

We now want to give an intrinsic definition of the tangent space. For a variety $V \subset \mathbb{A}^n$, $x \in V$, let $M_x = \mathcal{I}(x)$ be the maximal ideal of $R = C[V]$ vanishing at x. We have $C[V]/M_x \simeq C$; hence M_x/M_x^2 is a C-vector space (finite dimensional since M_x is a finitely generated R-module). Now $d_x f$, for arbitrary $f \in C[X_1, \ldots, X_n]$ can be viewed as a linear function on $\mathbb{A}^n$

(taking x as the origin), hence as a linear function on the vector subspace $\mathrm{Tan}(V)_x$ of $\mathbb{A}^n$. Since, for $f \in \mathcal{I}(V)$, $d_x f$ vanishes on $\mathrm{Tan}(V)_x$, $d_x f$ is determined on $\mathrm{Tan}(V)_x$ by the image of f in $C[V] = C[X_1, \dots, X_n]/\mathcal{I}(V)$. We then obtain a C-linear map d_x from R to the dual space of $\mathrm{Tan}(V)_x$. Since $R = C \oplus M_x$ as C vector spaces and $d_x(C) = 0$, we may view d_x as a map from M_x to the dual space of $\mathrm{Tan}(V)_x$.

Proposition 2.2.33. *The map d_x defines an isomorphism from M_x/M_x^2 to the dual space of* $\mathrm{Tan}(V)_x$.

Proof. The map is surjective because a linear function g on $\mathrm{Tan}(V)_x$ is the restriction of a linear function on $\mathbb{A}^n$ (with the origin at x) given by a linear polynomial $f(X_1, \dots, X_n)$ such that $d_x f$ is the given g. We now prove that $\mathrm{Ker}\, d_x = M_x^2$. Suppose $d_x f, f \in M_x$, vanishes on $\mathrm{Tan}(V)_x$, f being the image of some nonconstant $\widetilde{f} \in C[X_1, \dots, X_n]$. We have $d_x f = \sum a_i d_x f_i$, for some $a_i \in C, f_i \in \mathcal{I}(V)$. Setting $g = \widetilde{f} - \sum a_i d_x \widetilde{f_i}$, we see that $d_x g$ vanishes on all of $\mathbb{A}^n$; hence it is identically 0. Since $\widetilde{f}$ was nonconstant, we may assume that g is also. Then g must contain no linear term, i.e. g belongs to the square of the ideal $(X_1, \dots, X_n)$. The image of this ideal in R is M_x, and g has the same image f as $\widetilde{f}$ in R, so we conclude $f \in M_x^2$. □

We can now pass to the local ring $(\mathcal{O}_x, \mathfrak{M}_x)$, since $\mathcal{O}_x = R_{M_x}$ and $\mathfrak{M}_x = M_x R_{M_x}$. We obtain an isomorphism of C-vector spaces between the tangent space $\mathrm{Tan}(X)_x$ and the dual vector space of $\mathfrak{M}_x/\mathfrak{M}_x^2$ over C. For an algebraic variety V, we define the tangent space of V at a point x as $(\mathfrak{M}_x/\mathfrak{M}_x^2)^*$.

If X, Y are algebraic varieties, $x \in X$, a morphism of varieties $\varphi : X \to Y$ induces a C-algebra morphism $\varphi^* : \mathcal{O}_Y(U) \to \mathcal{O}_X(\varphi^{-1}(U))$, for U an open subset of Y containing $\varphi(x)$. It is clear that φ^* sends $\mathfrak{M}_{\varphi(x)}$ to $\mathfrak{M}_x$ and $\mathfrak{M}^2_{\varphi(x)}$ to $\mathfrak{M}_x^2$, hence defines a C-algebra morphism from $\mathfrak{M}_{\varphi(x)}/\mathfrak{M}^2_{\varphi(x)}$ to $\mathfrak{M}_x/\mathfrak{M}_x^2$. Considering the dual morphism, we obtain a morphism denoted $d_x\varphi$ and called *the differential mapping* of φ at the point x, from $\mathrm{Tan}(X)_x$ to $\mathrm{Tan}(Y)_{\varphi(x)}$.

Proposition 2.2.34. *Let $\varphi : X \to Y$ be an isomorphism of varieties, $x \in X$. Then* $\mathrm{Tan}(X)_x$ *is isomorphic to* $\mathrm{Tan}(Y)_{\varphi(x)}$.

Proof. It is clear that $d_{\varphi(x)}\varphi^{-1}$ is the inverse of $d_x\varphi$, so the two tangent spaces are isomorphic. □

We now look at the dimension of the tangent space. If $X \subset \mathbb{A}^n$ is an affine variety and $\mathcal{I}(X)$ is generated by $f_1, \dots, f_N$, the tangent space $\mathrm{Tan}(X)_x$ at a point x of X is defined by the equations

$$\sum_{j=1}^{n} (\partial f_i/\partial X_j)(x)(X_j - x_j) = 0,\ 1 \leq i \leq N.$$

Hence the dimension of $\mathrm{Tan}(X)_x$ is $n - r$ for r the rank of the matrix $((\partial f_i/\partial X_j)(x))_{1 \leq i \leq N, 1 \leq j \leq n}$. We now look at the dimension of the tangent space as x varies in X. Let ρ be the rank of the matrix $(\partial f_i/\partial X_j)_{1 \leq i \leq N, 1 \leq j \leq n}$ with entries in $C[X_1, \dots, X_n]$. Then all minors of order greater that ρ vanish and there exist non-zero minors Δ_k of order ρ. Hence in the matrix $((\partial f_i/\partial X_j)(x))$ all minors of order $> \rho$ vanish, so $r \leq \rho$ and the points for which $r < \rho$ are precisely those for which $\Delta_k(x) = 0$ for all k. Therefore there is a minimum value s for $\dim Tan(X)_x, x \in X$, and the points $x \in X$ for which $\dim Tan(X)_x > s$ form a proper closed subset of X.

We shall see in Proposition 2.2.36 below that, for X an irreducible variety, this minimum value s of $\dim Tan(X)_x$ is equal to $\dim X$. We say that x is a *simple point* or regular point or nonsingular point if $\dim Tan(X)_x = \dim X$. Otherwise we say that x is a singular point. A variety is called *nonsingular* or *smooth* if all its points are simple. Otherwise it is called singular. It is clear that $\mathbb{A}^n$ and $\mathbb{P}^n$ are smooth varieties. Exercise 9 gives examples of smooth and singular plane cubic curves.

Proposition 2.2.35. *Let V be a hypersurface in $\mathbb{A}^n$. Then* $\dim \mathrm{Tan}(V)_x \geq n-1$ *for all $x \in V$ and equality holds in a nonempty open subset of V.*

Proof. We have $V = \mathcal{V}(f)$, for some nonconstant $f \in C[X_1, \dots, X_n]$ and $\mathrm{Tan}(V)_x$ is defined by $d_x f = \sum_{i=1}^{n} (\partial f/\partial X_i)(x)(X_i - x_i) = 0$. We then have $\dim \mathrm{Tan}(V)_x = n - 1$ unless $(\partial f/\partial X_i)(x) = 0$ for all $i = 1, \dots, n$ and this condition determines a proper closed subset in V. □

Proposition 2.2.36. *Let X be an irreducible algebraic variety, $x \in X$. Then* $\dim \mathrm{Tan}(X)_x \geq \dim X$ *and equality holds in a nonempty open subset of X.*

Proof. As the statement is local, it is enough to consider the case of an affine variety. By Proposition 1.1.27, there exists a birational equivalence $\varphi : X \to Y$, where Y is a hypersurface in the affine space $\mathbb{A}^n$, where $n = 1 + \dim X$. According to Proposition 1.1.26, there exist nonempty open subsets $U \subset X, V \subset Y$ such that φ determines an isomorphism between them. By Proposition 2.2.35, the set W of regular points of Y is open and $\dim \mathrm{Tan}(Y)_y = n - 1$, for $y \in W$. The set $V \cap W$ is also open and so is $\varphi^{-1}(V \cap W) \subset U$. Since the dimension of the tangent space is invariant by isomorphisms (Proposition 2.2.34), $\dim \mathrm{Tan}(X)_x = \dim X$, for $x \in \varphi^{-1}(V \cap W)$. □

Remark 2.2.37. In the case of reducible varieties, the last result is no longer true. If, for example, $X = X_1 \cup X_2$, with $\dim X_1 = 1, \dim X_2 = 2$, and x is a simple point in X_1, not in X_2, we will have $\dim \mathrm{Tan}(X)_x = 1$, whereas $\dim X = 2$.

We may define the dimension of the variety X at a point x, denoted by $\dim_x X$ as the maximum of the dimensions of the irreducible components of X through x. We then say that a point x of X is nonsingular if $\dim T_x X = \dim_x X$.

Proposition 2.2.38. *Let $\varphi : X \to Y$ be a dominant morphism of irreducible varieties. Then there exists a nonempty open subset U of X such that for all $x \in U$, $\varphi(x)$ is a nonsingular point in Y and $d_x\varphi$ is a surjection from $T_x X$ onto $T_{\varphi(x)}Y$.*

Proof. By Proposition 2.2.36 we can take a nonempty open subset $U \subset Y$ of regular points of Y such that $\varphi^{-1}(U)$ is open and dense in X and $\varphi_{|\varphi^{-1}(U)} : \varphi^{-1}(U) \to U$ is regular and surjective. Hence φ induces a C-algebra monomorphism $\varphi^* : \mathcal{O}_Y(U) \to \mathcal{O}_X(\varphi^{-1}(U))$. As φ^* sends $\mathfrak{M}_{\varphi(x)}$ to $\mathfrak{M}_x$ and $\mathfrak{M}^2_{\varphi(x)}$ to $\mathfrak{M}^2_x$, we have the following commutative diagram

$$\begin{array}{ccc} \mathfrak{M}_{\varphi(x)} & \xrightarrow{\varphi^*} & \mathfrak{M}_x \\ \downarrow & & \downarrow \\ \mathfrak{M}_{\varphi(x)}/\mathfrak{M}^2_{\varphi(x)} & \xrightarrow{\overline{\varphi^*}} & \mathfrak{M}_x/\mathfrak{M}^2_x, \end{array}$$

where the upper horizontal arrow is injective. The statement on the surjectivity of $d_x\varphi$ is a consequence of the following lemma. (See [**Sh**] II6.2 Lemma 2.) □

Lemma 2.2.39. *With the hypothesis in Proposition 2.2.38, there exists a nonempty open subset $V \subset X$ such that $d_x\varphi$ is surjective for $x \in V$.*

Proof. The surjectivity of $d_x\varphi : T_x V \to T_{\varphi(x)}Y$ is dual to the injectivity of $\overline{\varphi^*} : \mathfrak{M}_{\varphi(x)}/\mathfrak{M}^2_{\varphi(x)} \to \mathfrak{M}_x/\mathfrak{M}^2_x$. We shall prove that, if $u_1, \dots, u_m$ are local parameters at $\varphi(x) \in Y$, i.e. $u_i \in \mathfrak{M}_{\varphi(x)}$ and their images form a basis of the vector space $\mathfrak{M}_{\varphi(x)}/\mathfrak{M}^2_{\varphi(x)}$, then $d_x(\varphi^*(u_1)), \dots, d_x(\varphi^*(u_m))$ are linearly independent. Considering the inclusion of $\mathcal{O}_{\varphi(x)}$ into the power series ring, it is visible that $u_1, \dots, u_m$ are algebraically independent and since $\varphi(X)$ is dense in Y, it follows that $u_1 \circ \varphi, \dots, u_m \circ \varphi$ are also algebraically independent as rational functions on X. We complete them to a system $v_1 = u_1 \circ \varphi, \dots, v_m = u_m \circ \varphi, v_{m+1}, \dots, v_n$ of $n = \dim X$ algebraically independent functions.

The lemma will be proved if we check that for any system $w_1, \dots, w_n$ of algebraically independent functions on X, the set of points z at which $w_1, \dots, w_n$ are local parameters for $\mathcal{O}_z$ is open and nonempty. We assume that $X \subset \mathbb{A}^N$, with coordinates $x_1, \dots, x_N$. We prove that for points z of a nonempty open subset $U \subset X$ all differentials $d_z x_i$ can be expressed as linear combinations of $d_z w_1, \dots, d_z w_n$. If these were linearly dependent, it would then follow that $\dim T_z X < n$. Now, each x_i is related to $w_1, \dots, w_n$ by a relation

$$F_i(x_i, w_1, \dots, w_n) = 0, \tag{2.1}$$

with F_i an irreducible polynomial and hence, as $\operatorname{char} C = 0$, $\dfrac{\partial F_i}{\partial x_i} \neq 0$. Suppose that

$$F_i = a_0 x_i^{n_i} + a_1 x_i^{n_i - 1} + \cdots + a_{n_i},$$

with $a_j \in C[w_1, \dots, w_n]$. Now $d_x a_j$ are linear combinations of $d_x w_1, \dots, d_x w_n$. Using the Leibniz rule of the differential d_x, it follows from (2.1) that

$$\frac{\partial F_i}{\partial x_i}(z) d_z x_i + x_i^{n_i} d_z a_0 + \cdots + d_z a_{n_i} = 0$$

at any point $z \in X$. The points at which all $\dfrac{\partial F_i}{\partial x_i}(z) \neq 0$ form a nonempty open set in X and at such points $d_z x_i, i = 1, \dots, N$, can be presented as linear combinations of $d_z w_1, \dots, d_z w_n$. Hence the lemma is proved. □

We shall now introduce the notion of completeness for varieties.

Definition 2.2.40. A variety X is called *complete* if for all varieties Y, $pr_2 : X \times Y \to Y$ is a closed map (i.e. sends closed subsets to closed subsets).

Evidently a single point, viewed as a variety, is complete. It is also clear that X is complete if and only if all its irreducible components are. Clearly, for a variety X to be complete it suffices that it satisfies the condition in the definition for varieties Y irreducible and affine. The affine line $\mathbb{A}^1$ is not complete as $\mathcal{V}(X_1 X_2 - 1) \subset \mathbb{A}^1 \times \mathbb{A}^1$ projects to a nonclosed subset of $\mathbb{A}^1$.

Remark 2.2.41. The analogous property for topological spaces of the condition in the definition of complete variety is a characterization of compact spaces. More precisely, a theorem of Kuratowski gives that for a Hausdorff space X compactness is equivalent to the following property: the projection $pr_2 : X \times Y \to Y$ is a closed map for any topological space Y, with

$X \times Y$ endowed with the usual product topology. (See [**E**] 3.1.16.) Hence completeness of varieties can be understood as an analogue of compactness of topological spaces.

The next proposition gives the first properties of completeness. We leave its proof as an exercise. (See Exercise 17.)

Proposition 2.2.42. *Let X, Y be varieties.*

a) If X is complete and Y is closed in X, then Y is complete.

b) If X and Y are complete, then $X \times Y$ is complete.

c) If $\varphi : X \to Y$ is a morphism and X is complete, then $\varphi(X)$ is closed and complete.

d) If Y is a complete subvariety of X, then Y is closed.

e) If X is complete and affine, then $\dim X = 0$.

f) A complete quasiprojective variety is projective.

Proposition 2.2.43. *Any projective variety is complete.*

Proof. By Proposition 2.2.42 a), it is enough to prove that $pr_2 : \mathbb{P}^n \times Y \to Y$ is closed for any variety Y. We may assume that Y is affine and irreducible; let $R = C[Y]$. Let $U_i = \{(x_0 : x_1 : \cdots : x_n) \in \mathbb{P}^n : x_i \neq 0\}$ be the affine open sets covering $\mathbb{P}^n$. The affine open sets $V_i := U_i \times Y$ cover $\mathbb{P}^n \times Y$. If $X_0, X_1, \ldots, X_n$ are homogeneous coordinates in $\mathbb{P}^n$, then $C[V_i]$ can be described as $R[X_0/X_i, \ldots, X_n/X_i]$. Take any closed set Z in $\mathbb{P}^n \times Y$ and any point $y \in Y \setminus pr_2(Z)$. We want to find a neighborhood of y in Y of the form Y_f which is disjoint from $pr_2(Z)$. This amounts to finding $f \in R, f \notin M = \mathcal{I}(f)$ such that f vanishes on $pr_2(Z)$, i.e. such that $j_i^*(pr_2^*(f)) \in \mathcal{I}(Z_i)$, for $j_i : V_i \hookrightarrow \mathbb{P}^n \times Y$, $Z_i = Z \cap V_i$. The existence of such f will follow from Nakayama's lemma applied to a suitable R-module, which we now proceed to construct.

We first consider the polynomial ring $S = R[X_0, \ldots, X_n]$, with the natural grading $S = \sum S_m$. We construct a homogeneous ideal $I \subset S$ by letting I_m consist of all $f(X_0, \ldots, X_n) \in S_m$ such that $f(X_0/X_i, \ldots, X_n/X_i) \in \mathcal{I}(Z_i)$ for all i.

Let $f \in \mathcal{I}(Z_i)$, for i fixed. We claim that multiplication by a sufficiently high power of X_i will take f into I. Indeed, if we view f as a polynomial in $X_0/X_i, \ldots, X_n/X_i$, then $X_i^m f$ becomes a homogeneous polynomial of degree m in $X_0, \ldots, X_m$ for large m. Now $(X_i^m/X_j^m)f \in R_j$ vanishes on $Z_i \cap V_j = Z_j \cap V_i$ while $(X_i^{m+1}/X_j^{m+1})f$ vanishes at all points of Z_j not in V_i. Since j is arbitrary, we conclude that $X_i^{m+1} f$ lies in I_{m+1}.

Now Z_i and $U_i \times \{y\}$ are disjoint closed subsets of the affine variety V_i, so their ideals $\mathcal{I}(Z_i)$ and MR_i together generate the unit ideal R_i. In particular, we have an equality $1 = f_i + \sum_j m_{ij} g_{ij}$, where $f_i \in \mathcal{I}(Z_i), m_{ij} \in M, g_{ij} \in R_i$. We have seen that multiplication by a sufficiently high power of X_i takes $f \in \mathcal{I}(Z_i)$ into I. We can choose it large enough so that it works for all i and moreover it takes g_{ij} into S. So we obtain $X_i^m \in I_m + MS_m$, for all i. Enlarging m even more, we can get all monomials of degree m in $X_0, \ldots, X_n$ to lie in $I_m + MS_m$. This implies that $S_m = I_m + MS_m$. We can now apply Nakayama's lemma ([**Ma**] 1.M) to the finitely generated R-module S_m/I_m which satisfies $M(S_m/I_m) = S_m/I_m$ and obtain that there exists $f \in R, f \notin M$ such that $f(S_m/I_m) = 0$, thus $fS_m \subset I_m$. In particular, $fX_i^m \in I_m$, so f vanishes on $pr_2(Z)$. □

Exercises

(1) Let $X = \{(x,y) \in \mathbb{A}^2 : x^2 = y^3\}$. Define $\varphi : X \to \mathbb{A}^1$ by $\varphi(x,y) = xy^{-1}$ if $(x,y) \neq (0,0)$ and $\varphi(0,0) = 0$. Show that φ is a morphism of irreducible varieties which is birational and bijective but is not an isomorphism of varieties.

(2) For X, Y irreducible varieties, prove $\dim(X \times Y) = \dim X + \dim Y$.

(3) *Product of projective varieties.* If $X \subset \mathbb{P}^n, Y \subset \mathbb{P}^m$ are projective varieties the cartesian product cannot be straightforwardly identified with a subset of $\mathbb{P}^n \times \mathbb{P}^m$. We consider the map $\varphi : \mathbb{P}^n \times \mathbb{P}^m \to \mathbb{P}^q$, where $q = (n+1)(m+1) - 1$, defined by

$$\varphi((x_0,\dots,x_n),(y_0,\dots,y_m)) = (x_0y_0,\dots,x_0y_m,x_1y_0,\dots,x_1y_m,\dots,x_ny_0,\dots,x_ny_m).$$

a) Prove that φ is well defined and its image is closed in $\mathbb{P}^q$. Prove that if X is closed in $\mathbb{P}^n$ and Y is closed in $\mathbb{P}^m$, then $\varphi(X \times Y)$ is closed in $\mathbb{P}^q$.
Hint: Use the covering of the projective space by affine open sets.

b) Prove that the construction of $X \times Y$ as a closed set of $\mathbb{P}^q$ is compatible with the product of prevarieties defined in Proposition 2.1.6.

(4) Exhibit a constructible subset of $\mathbb{A}^2$ not locally closed.

(5) Prove that constructible subsets of a topological space X form the boolean algebra generated by the open (resp. closed) subsets of X, i.e. the smallest collection containing all open (resp. closed) subsets which is closed under intersections, finite unions, and complements.

(6) Let V be an irreducible affine variety for which $C[V]$ is a unique factorization domain. Prove that each closed subset W of codimension 1 has the form $\mathcal{V}(f)$ for some $f \in C[V]$.
Hint: First treat the case W irreducible. Show that minimal prime ideals of $C[V]$ are principal.

(7) Let $X \subset \mathbb{A}^n$ be an affine variety. If $f_1,\dots,f_r$ generate $\mathcal{I}(X)$, prove that $d_xf_1,\dots,d_xf_r$ generate $\mathcal{I}(\mathrm{Tan}(X)_x)$, for $x \in X$.

(8) Prove that a conic, i.e. a plane algebraic curve of degree two, is smooth if and only if it is nondegenerate.

(9) Consider the cubic $\mathcal{C} = \mathcal{V}(Y^2 - X^3 - aX - b) \subset \mathbb{A}^3$. Prove that $\mathcal{C}$ has a singular point if and only if $\mathrm{disc}\,(X^3 + aX + b) = 0$. Prove that $\mathcal{C}$ has at most one singular point.

Consider the projective closure $\overline{\mathcal{C}}$ of $\mathcal{C}$. Prove that it has a single point P at infinity, which is nonsingular and that $\mathrm{Tan}_P(\mathcal{C})$ is the line at infinity.

(10) a) Consider the cubic $\mathcal{C} = \mathcal{V}(Y^2 - (X-a)^2(X-b)) \subset \mathbb{A}^3$, with $a \neq b$. Find the singular point P of $\mathcal{C}$. Show that there are exactly two lines through P such that P is a triple solution of the intersection of the line with $\mathcal{C}$. In this case, we say that P is a *node* of $\mathcal{C}$. Make a change of variables (over $\mathbb{C}$) taking the equation of $\mathcal{C}$ to $Y^2 = X^3 - X^2$ and draw this curve (over $\mathbb{R}$).

b) Consider the cubic $\mathcal{C} = \mathcal{V}(Y^2 - (X-a)^3) \subset \mathbb{A}^3$. Find the singular point P of $\mathcal{C}$. Show that there is exactly one line through P such that P is a triple solution of the intersection of the line with $\mathcal{C}$. In this case, we say that P is a *cusp* of $\mathcal{C}$. Make a change of variables taking the equation of $\mathcal{C}$ to $Y^2 = X^3$ and draw this curve.

(11) Let us denote by $\mathrm{Sing}\, V$ the set of singular points of a variety V. Let V be an algebraic variety, $V_i, 1 \leq i \leq n$, its irreducible components. Prove that

$$\mathrm{Sing}\, V = \bigcup_{i=1}^{n} \mathrm{Sing}\, V_i \cup \bigcup_{i \neq j} (V_i \cap V_j).$$

(12) Let W be a subvariety of a variety V, $j : W \to V$ the canonical injection and $x \in W$. Prove that the map $d_x j : T_x(W) \to T_x(V)$ is injective.

(13) Let X, Y be algebraic varieties, $x \in X, y \in Y$. Prove

$$\mathrm{Tan}(X \times Y)_{(x,y)} \simeq \mathrm{Tan}(X)_x \oplus \mathrm{Tan}(Y)_y.$$

(14) *The tangent cone.* If $f \in C[X_1, \dots, X_n]$, we consider the Taylor expansion $f = f_0 + f_1 + \dots + f_d$ of f at $x = (x_1, \dots, x_n) \in C^n$, where

$$f_k = \sum_{i_1 + \dots + i_n = k} \frac{1}{i_1! \dots i_n!} \frac{\partial^k f}{\partial x_1^{i_1} \dots \partial x_n^{i_n}}(x)(X_1 - x_1)^{i_1} \dots (X_n - x_n)^{i_n}$$

and d is the total degree of f. We denote by $in(f)$ the initial form of f, i.e. the homogeneous polynomial f_k such that $f_0 = \dots = f_{k-1} = 0$ and $f_k \neq 0$. If V is an affine variety in $\mathbb{A}^n$, $x \in V$, we define the tangent cone $TC(V)_x$ of V at x by $TC(V)_x := \mathcal{V}(\{in(f) : f \in \mathcal{I}(V)\})$. It is clear that $TC(V)_x = \mathrm{Tan}(V)_x$ if x is a simple point of V.

a) Prove that $TC(V)_x$ is a cone with vertex in x, i.e. that if $y \in TC(V)_x$, the line joining x to y is contained in $TC(V)_x$.

b) Compute the tangent cones of the cubic curves in Exercise 10 at the singular points.

(15) a) Let R be a noetherian local ring, with unique maximal ideal $\mathfrak{M}$. Prove that $\mathfrak{M}$ is generated as an R-module by $f_1, \ldots, f_n \Leftrightarrow \mathfrak{M}/\mathfrak{M}^2$ is generated as an $R/\mathfrak{M}$-module by the images of $f_1, \ldots, f_n$.

b) Recall that a local ring $(R, \mathfrak{M})$ is called *regular* if its Krull dimension coincides with the minimal number of generators of $\mathfrak{M}$. Prove that for x a simple point of an algebraic variety X, the local ring $\mathcal{O}_x$ is a regular local ring, hence is integrally closed in $C(X)$.

Hint: See [**A-M**], *chap. 11.*

(16) Let V be an n-dimensional C-vector space, with exterior algebra ΛV. To a subspace W of V of dimension d, we associate the point in $\mathbb{P}(\Lambda^d V)$ corresponding to $v_1 \wedge \cdots \wedge v_d$, for $(v_1, \ldots, v_d)$ a basis of W. Let φ denote the map obtained in this way from the collection $\Sigma_d(V)$ of subspaces of V of dimension d to the projective space $\mathbb{P}(\Lambda^d V)$.

a) Prove that φ is well defined and its image is a closed subset of $\mathbb{P}(\Lambda^d V)$. This projective variety is called *Grassmann variety*.

b) Write down the equations for the Grassmann variety of lines in $\mathbb{A}^3$.

(17) Provide the proof of Proposition 2.2.42.

Hint: For c) use that the graph of φ is closed in $X \times Y$ (Proposition 2.2.8). For e) apply c) to morphisms $X \to \mathbb{A}^1$ and use that $\mathbb{A}^1$ is not complete.

(18) To an element $A \in \mathrm{GL}(2, C)$, i.e.

$$A = \begin{pmatrix} a & b \\ c & d \end{pmatrix}, \; a, b, c, d \in C, \; \det A = ad - bc \neq 0,$$

we associate a map

$$\begin{array}{cccc} \varphi_A : & \mathbb{P}^1 & \to & \mathbb{P}^1 \\ & (x : y) & \mapsto & (ax + by : cx + dy). \end{array}$$

a) Prove that φ_A is an automorphism of the variety $\mathbb{P}^1$ and φ_A is the identity if and only if A is a scalar matrix.

b) Let $\mathrm{PGL}(2, C)$ be the quotient group of $\mathrm{GL}(2, C)$ by the subgroup of scalar matrices. Prove that the map

$$\begin{array}{ccc} \mathrm{PGL}(2, C) & \to & Aut\, \mathbb{P}^1 \\ \overline{A} & \mapsto & \varphi_A \end{array}$$

is a group isomorphism.

Hint: Prove that an automorphism of $\mathbb{P}^1$ is determined by the images of the three points $0 = (0 : 1), 1 = (1 : 1), \infty = (1 : 0)$.

Part 2

Algebraic Groups

In Part 2, we introduce algebraic groups which are algebraic varieties endowed with a compatible group structure. We are mainly interested in linear algebraic groups, that is, closed subgroups of some general linear group, as these appear as differential Galois groups of linear homogeneous differential equations. We shall see that, for an algebraic group, being a linear algebraic group and being an affine variety is equivalent. The fact that the quotient of an algebraic group by a closed subgroup is an algebraic group will be used in the fundamental theorem of Picard-Vessiot theory. In chapter 4, we introduce the notion of Lie algebra and define the Lie algebra associated to a linear algebraic group. The concept of solvability and the Lie-Kolchin theorem will be used in the characterization of linear differential equations solvable by quadratures. We present the classification of the closed subgroups of the special linear group of degree 2 and give a geometric proof for the determination of the finite ones.

Throughout Part 2, C will denote an algebraically closed field of characteristic 0.

Chapter 3

Basic Notions

In this chapter we define and provide examples of algebraic groups. We give the notion of a connected algebraic group. We see that a linear algebraic group is an affine variety and reciprocally that an affine algebraic group is a closed subgroup of some general linear group. We consider actions of algebraic groups on algebraic varieties. We end the chapter with the construction of the quotient of an algebraic group by a normal closed subgroup.

3.1. The notion of algebraic group

Definition 3.1.1. An *algebraic group* over C is an algebraic variety G defined over C, endowed with the structure of a group and such that the two maps $\mu : G \times G \to G$, where $\mu(x, y) = xy$ and $\iota : G \to G$, where $\iota(x) = x^{-1}$, are morphisms of varieties.

Translation by an element $y \in G$, i.e. $x \mapsto xy$ is clearly a variety automorphism of G, and therefore all geometric properties at one point of G can be transferred to any other point, by suitable choice of y. For example, since G has simple points (Proposition 2.2.36), all points must be simple; hence G is nonsingular.

Example 3.1.2. The *additive group* $\mathbb{G}_a$ is the affine line $\mathbb{A}^1$ with the group law $\mu(x, y) = x + y$, so $\iota(x) = -x$ and $e = 0$. The *multiplicative group* $\mathbb{G}_m$ is the principal open set $C^* \subset \mathbb{A}^1$ with group law $\mu(x, y) = xy$, so $\iota(x) = x^{-1}$ and $e = 1$. Each of these two groups is irreducible, as a variety, and has dimension 1. It can be proven that they are the only (up to isomorphism) affine algebraic groups with these two properties. (See Exercise 11 in chapter 4.)

Example 3.1.3. The *general linear group* $\mathrm{GL}(n, C)$ is the group of all invertible $n \times n$ matrices with entries in C with matrix multiplication. As already noted, the set $M(n, C)$ of all $n \times n$ matrices over C may be identified with the affine space of dimension n^2 and $\mathrm{GL}(n, C)$ with the principal open subset defined by the non-vanishing of the determinant. Viewed thus as an affine variety, $\mathrm{GL}(n, C)$ has a coordinate ring generated by the restriction of the n^2 coordinate functions X_{ij}, together with $1/\det(X_{ij})$. The formulas for matrix multiplication and inversion make it clear that $\mathrm{GL}(n, C)$ is an algebraic group. Notice that $\mathrm{GL}(1, C) = \mathbb{G}_m$.

If V is a finite dimensional C-vector space we define $\mathrm{GL}(V)$ as the group of C-vector space automorphisms of V. If $n = \dim_C V$, we have $\mathrm{GL}(V) \simeq \mathrm{GL}(n, C)$.

Example 3.1.4. Taking into account that a closed subgroup of an algebraic group is again an algebraic group, we can construct further examples. We consider the following subgroups of $\mathrm{GL}(n, C)$:

(1) $\mathrm{SL}(n, C) := \{A \in \mathrm{GL}(n, C) : \det A = 1\}$ (*special linear group*);
(2) $\mathrm{T}(n, C) := \{(a_{ij}) \in \mathrm{GL}(n, C) : a_{ij} = 0, i > j\}$ (*upper triangular group*);
(3) $\mathrm{U}(n, C) := \{(a_{ij}) \in \mathrm{GL}(n, C) : a_{ii} = 1, a_{ij} = 0, i > j\}$ (*upper triangular unipotent group*);
(4) $\mathrm{D}(n, C) := \{(a_{ij}) \in \mathrm{GL}(n, C) : a_{ij} = 0, i \neq j\}$ (*diagonal group*).

A *linear algebraic group* is a closed subgroup of some $\mathrm{GL}(n, C)$. The groups above are then examples of linear algebraic groups.

Example 3.1.5. The *direct product* of two or more algebraic groups, i.e. the usual direct product of groups endowed with the algebraic variety structure of the product (see Proposition 2.1.6 and Example 2.2.4), is again an algebraic group. For example $D(n, C)$ may be viewed as the direct product of n copies of $\mathbb{G}_m$, while affine n-space may be viewed as the direct product of n copies of $\mathbb{G}_a$.

3.2. Connected algebraic groups

Let G be an algebraic group. We assert that only one irreducible component of G contains the unit element e. Indeed, let $X_1, \dots, X_m$ be the distinct irreducible components containing e. The image of the irreducible variety $X_1 \times \cdots \times X_m$ under the product morphism is an irreducible subset $X_1 \cdots X_m$ of G which again contains e. So $X_1 \cdots X_m$ lies in some X_i. On the other hand, each of the components $X_1, \dots, X_m$ clearly lies in $X_1 \cdots X_m$. Then m must be 1.

We denote by G^0 this unique irreducible component containing e and call it the *identity component of* G.

Proposition 3.2.1. *Let G be an algebraic group.*

a) G^0 is a normal subgroup of finite index in G, whose cosets are the connected as well as irreducible components of G.

b) Each closed subgroup of finite index in G contains G^0.

c) Every finite conjugacy class of G has at most as many elements as $[G : G^0]$.

Proof. a) For each $x \in G^0$, $x^{-1}G^0$ is an irreducible component of G containing e, so $x^{-1}G^0 = G^0$. Therefore $G^0 = (G^0)^{-1}$, and further $G^0G^0 = G^0$, i.e. G^0 is a (closed) subgroup of G. For any $x \in G$, xG^0x^{-1} is also an irreducible component of G containing e, so $xG^0x^{-1} = G^0$ and G^0 is normal. Its (left or right) cosets are translates of G^0, and so must also be irreducible components of G; as G is a Noetherian space there can only be finitely many of them. Since they are disjoint, they are also the connected components of G.

b) If H is a closed subgroup of finite index in G, then each of its finitely many cosets is also closed. The union of those cosets distinct from H is also closed and then, H is open. Therefore the left cosets of H give a partition of G^0 into a finite union of open sets. Since G^0 is connected and meets H, we get $G^0 \subset H$.

c) Write $n = [G : G^0]$ and assume that there exists an element $x \in G$ with a finite conjugacy class having a number of elements exceeding n. The mapping from G to G defined by $a \mapsto axa^{-1}$ is continuous. The inverse image of each conjugate of x is closed and, as there are finitely many of them, also open. This yields a decomposition of G into more than n open and closed sets, a contradiction. □

We shall call an algebraic group G *connected* when $G = G^0$. As usual in the theory of linear algebraic groups, we shall reserve the word "irreducible" for group representations.

The additive group $\mathbb{G}_a(C)$ and the multiplicative group $\mathbb{G}_m(C)$ are connected groups. The group $\mathrm{GL}(n, C)$ is connected, as it is a principal open set in the affine space of dimension n^2. The next proposition will allow us to deduce the connectedness of some other algebraic groups. We first establish the following lemma.

Lemma 3.2.2. *Let U, V be two dense open subsets of an algebraic group G. Then $G = U \cdot V$.*

Proof. Since inversion is a homeomorphism, V^{-1} is again a dense open set. So is its translate xV^{-1}, for any given $x \in G$. Therefore, U must meet xV^{-1}, forcing $x \in U \cdot V$. $\square$

For an arbitrary subset M of an algebraic group G, we define the *group closure* $\mathrm{GC}(M)$ of M as the intersection of all closed subgroups of G containing M.

Proposition 3.2.3. *Let G be an algebraic group, $f_i : X_i \to G$, $i \in I$, a family of morphisms from irreducible varieties X_i to G, such that $e \in Y_i = f_i(X_i)$ for each $i \in I$. Set $M = \cup_{i \in I} Y_i$. Then*

a) $\mathrm{GC}(M)$ is a connected subgroup of G.

b) For some finite sequence $a = (a_1, \ldots, a_n)$ in I, $GC(M) = Y_{a_1}^{e_1} \ldots Y_{a_n}^{e_n}$, $e_i = \pm 1$.

Proof. We can if necessary enlarge I to include the morphisms $x \mapsto f_i(x)^{-1}$ from X_i to G. For each finite sequence $a = (a_1, \ldots, a_n)$ in I, set $Y_a := Y_{a_1} \ldots Y_{a_n}$. The set Y_a is constructible, as it is the image of the irreducible variety $X_{a_1} \times \cdots \times X_{a_n}$ under the morphism $f_{a_1} \times \cdots \times f_{a_n}$ composed with multiplication in G, and moreover $\overline{Y_a}$ is an irreducible variety passing through e (since closure and homeomorphic image of an irreducible set are so). Given two finite sequences b, c in I, we have $\overline{Y_b}\,\overline{Y_c} \subset \overline{Y_{(b,c)}}$, where (b, c) is the sequence obtained from b and c by juxtaposition. Indeed, for $x \in Y_c$, the map $y \mapsto yx$ sends Y_b into $Y_{(b,c)}$, hence by continuity $\overline{Y_b}$ into $\overline{Y_{(b,c)}}$, i.e. $\overline{Y_b} Y_c \subset \overline{Y_{(b,c)}}$. This last inclusion shows as well that multiplication by an element in $\overline{Y_b}$ sends Y_c into $\overline{Y_{(b,c)}}$, hence $\overline{Y_c}$ as well. Let us now take a sequence a for which $\overline{Y_a}$ is maximal. For each finite sequence b, we have $\overline{Y_a} \subset \overline{Y_a}\,\overline{Y_b} \subset \overline{Y_{(a,b)}} = \overline{Y_a}$. Setting $b = a$, we have $\overline{Y_a}$ stable under multiplication. Choosing b such that $Y_b = Y_a^{-1}$, we also have $\overline{Y_a}$ stable under inversion. We have then that $\overline{Y_a}$ is a closed subgroup of G containing all Y_i so $\overline{Y_a} = \mathrm{GC}(M)$, proving a).

Since Y_a is constructible, Lemma 3.2.2 shows that $\overline{Y_a} = Y_a \cdot Y_a = Y_{(a,a)}$, so the sequence (a, a) satisfies b). $\square$

Corollary 3.2.4. *Let G be an algebraic group, $Y_i, i \in I$, a family of closed connected subgroups of G which generate G as an abstract group. Then G is connected.* $\square$

Corollary 3.2.5. *The algebraic groups $\mathrm{SL}(n, C), \mathrm{U}(n, C), \mathrm{D}(n, C), \mathrm{T}(n, C)$ (see Example 3.1.4) are connected.*

Proof. Let U_{ij} be the group of all matrices with 1's on the diagonal, arbitrary entry in the (i, j) position and 0's elsewhere, for $1 \leq i, j \leq n, i \neq j$.

Then the U_{ij} are isomorphic to $\mathbb{G}_a(C)$, and so connected, and generate $\mathrm{SL}(n, C)$. Hence by Corollary 3.2.4, $\mathrm{SL}(n, C)$ is connected. The U_{ij} with $i < j$ generate $\mathrm{U}(n, C)$; whence $\mathrm{U}(n, C)$ is connected.

The group $\mathrm{D}(n, C)$ is the direct product of n copies of $\mathbb{G}_m(C)$, whence connected. Finally, $\mathrm{T}(n, C)$ is generated by $\mathrm{U}(n, C)$ and $\mathrm{D}(n, C)$; whence is also connected. □

3.3. Subgroups and morphisms

Lemma 3.3.1. *Let E be a constructible subset of a topological space X. Then E contains a dense open subset of its closure.*

Proof. We have $E = \cup_{i=1}^k U_i \cap V_i$, where U_i are open in X and V_i are closed in X. We can assume V_i irreducible (by substituting each V_i if necessary by the union of its irreducible components). Let $U_i^* := U_i \setminus (\cup_{j \neq i} V_j)$; clearly U_i^* is open in X. As V_i is irreducible, we have $\overline{V_i \cap U_i^*} = V_i$. Let us observe that $U_i^* \cap (\cup_{i=1}^k U_i \cap V_i) = U_i^* \cap V_i$. We take $U := \cup_{i=1}^k U_i^*$; then U is open in X. Now $U \cap E = U \cap (\cup_{i=1}^k U_i \cap V_i) = \cup_{i=1}^k U_i^* \cap V_i$. We have $\overline{U \cap E} = \cup_{i=1}^k \overline{U_i^* \cap V_i} = \cup_{i=1}^k V_i = \overline{E}$. □

Proposition 3.3.2. *Let H be a subgroup of an algebraic group G, $\overline{H}$ its closure.*

a) $\overline{H}$ is a subgroup of G.

b) If H is constructible, then $H = \overline{H}$.

Proof. a) Inversion being a homeomorphism, it is clear that $\overline{H}^{-1} = \overline{H^{-1}} = \overline{H}$. Similarly, translation by $x \in H$ is a homeomorphism, so $x\overline{H} = \overline{xH} = \overline{H}$, i.e. $H\overline{H} \subset \overline{H}$. In turn, if $x \in \overline{H}$, $Hx \subset \overline{H}$, so $\overline{H}x = \overline{Hx} \subset \overline{H}$. This says that $\overline{H}$ is a group.

b) If H is constructible, by Lemma 3.3.1, it contains a dense open subset U of $\overline{H}$. Since $\overline{H}$ is a group, by part a), Lemma 3.2.2 shows that $\overline{H} = U \cdot U \subset H \cdot H = H$. □

For a subgroup H of a group G we define the *normalizer* $N_G(H)$ *of* H *in* G as

$$N_G(H) = \{x \in G : xHx^{-1} = H\}.$$

If a subgroup H' of G is contained in $N_G(H)$, we say that H' *normalizes* H.

Proposition 3.3.3. *Let A, B be closed subgroups of an algebraic group G. If B normalizes A, then AB is a closed subgroup of G.*

Proof. Since $B \subset N_G(A)$, AB is a subgroup of G. Now AB is the image of $A \times B$ under the product morphism $G \times G \to G$; hence it is constructible by Theorem 2.2.21 and therefore closed by Proposition 3.3.2 b). □

By definition a *morphism of algebraic groups* is a group homomorphism which is also a morphism of algebraic varieties.

Proposition 3.3.4. *Let $\varphi : G \to G'$ be a morphism of algebraic groups. Then*

a) $\operatorname{Ker}\varphi$ *is a closed subgroup of* G.

b) $\operatorname{Im}\varphi$ *is a closed subgroup of* G'.

c) $\varphi(G^0) = \varphi(G)^0$.

d) $\dim G = \dim(\operatorname{Ker}\varphi) + \dim(\operatorname{Im}\varphi)$.

Proof. a) φ is continuous and $\operatorname{Ker}\varphi$ is the inverse image of the closed set $\{e\}$.

b) $\varphi(G)$ is a subgroup of G'. It is also a constructible subset of G', by Theorem 2.2.21 , so it is closed by Proposition 3.3.2 b).

c) $\varphi(G^0)$ is closed by b) and connected; hence it lies in $\varphi(G)^0$. As it has finite index in $\varphi(G)$, it must be equal to $\varphi(G)^0$, by Proposition 3.2.1b).

d) The fibres of the morphism $G \to \varphi(G)$ induced by φ are the cosets of G modulo $\operatorname{Ker}\varphi$; hence they all have dimension equal to $\dim(\operatorname{Ker}\varphi)$. So d) follows from Proposition 2.2.24. □

3.4. Linearization of affine algebraic groups

We have seen that any closed subgroup of $\mathrm{GL}(n, C)$ is an affine algebraic group. We shall now see that the converse is also true.

Let G be an algebraic group, V an affine variety. We say that V is a *G-variety* if the algebraic group G acts on the affine variety V, i.e. we have a morphism of algebraic varieties

$$\begin{array}{ccc} G \times V & \to & V \\ (x, v) & \mapsto & x.v \end{array}$$

satisfying $x_1.(x_2.v) = (x_1 x_2).v$, for any x_1, x_2 in G, v in V, and $e.v = v$, for any $v \in V$.

Let V, W be G-varieties. A morphism $\varphi : V \to W$ is a G-morphism, or is said to be *equivariant* if $\varphi(x.v) = x.\varphi(v)$, for $x \in G, v \in V$.

The action of G over V induces an action of G on the coordinate ring $C[V]$ of V defined by

$$\begin{array}{rcl} G \times C[V] & \to & C[V] \\ (x, f) & \mapsto & x.f : v \mapsto f(x^{-1}.v). \end{array}$$

In particular, we can consider two different actions of G on its coordinate ring $C[G]$ associated to the action of G on itself by left or right translations. To the action of G on itself by left translations defined by

$$\begin{array}{rcl} G \times G & \to & G \\ (x, y) & \mapsto & xy \end{array}$$

corresponds the action

$$\begin{array}{rcl} G \times C[G] & \to & C[G] \\ (x, f) & \mapsto & \lambda_x(f) : y \mapsto f(x^{-1}y). \end{array} \tag{3.1}$$

To the action of G on itself by right translations defined by

$$\begin{array}{rcl} G \times G & \to & G \\ (x, y) & \mapsto & yx^{-1} \end{array}$$

corresponds the action

$$\begin{array}{rcl} G \times C[G] & \to & C[G] \\ (x, f) & \mapsto & \rho_x(f) : y \mapsto f(yx). \end{array} \tag{3.2}$$

We can use right translations to characterize membership in a closed subgroup:

Lemma 3.4.1. *Let H be a closed subgroup of an algebraic group G, I the ideal of $C[G]$ vanishing on H. Then*

$$H = \{x \in G : \rho_x(I) \subset I\}.$$

Proof. Let $x \in H$. If $f \in I$, $\rho_x(f)(y) = f(yx) = 0$ for all $y \in H$; hence $\rho_x(f) \in I$, i.e. $\rho_x(I) \subset I$. Assume now $\rho_x(I) \subset I$. In particular, if $f \in I$, then $\rho_x(f)$ vanishes at $e \in H$, then $f(x) = f(ex) = \rho_x(f)(e) = 0$, so $x \in \mathcal{V}(\mathcal{I}(H)) = H$ (Exercise 4 in chapter 1), as the preceding equality holds for all $f \in I$. □

Lemma 3.4.2. *Let G be an algebraic group and V an affine variety both defined over the field C. Assume that G acts on V and let F be a finite dimensional C-vector subspace of the coordinate ring $C[V]$ of V.*

a) There exists a finite dimensional subspace E of $C[V]$ including F which is stable under the action of G.

b) F itself is stable under the action of G if and only if $\varphi^*F \subset C[G] \otimes_C F$, where $\varphi : G \times V \to V$ is given by $\varphi(x, v) = x^{-1}.v$

Proof. a) If we prove the result in the case in which F has dimension 1, we can obtain it for a finite dimensional F by summing up the subspaces E corresponding to the subspaces of F generated by one vector of a chosen basis of F. So we may assume that $F =< f >$ for some $f \in C[V]$. Let $\varphi : G \times V \to V$ be the morphism giving the action of G on V, $\varphi^* : C[V] \to C[G \times V] = C[G] \otimes C[V]$ the corresponding morphism between coordinate rings. Let us write $\varphi^* f = \sum g_i \otimes f_i \in C[G] \otimes C[V]$. (Note that this expression is not unique.) For $x \in G, v \in V$, we have $(x.f)(v) = f(x^{-1}.v) = f(\varphi(x^{-1}, v)) = (\varphi^* f)(x^{-1}, v) = \sum g_i(x^{-1}) f_i(v)$ and then $x.f = \sum g_i(x^{-1}) f_i$. So every translate $x.f$ of f is contained in the finite dimensional C-vector space of $C[V]$ generated by the functions f_i. So $E = \langle x.f \,|\, x \in G \rangle$ is a finite-dimensional G-stable vector space containing f.

b) If $\varphi^* F \subset C[G] \otimes_C F$, then the functions f_i in the proof of a) can be taken to lie in F; therefore F is stable under the action of G. Conversely, let F be stable under the action of G and extend a vector space basis $\{f_i\}$ of F to a basis $\{f_i\} \cup \{h_j\}$ of $C[V]$. If $\varphi^* f = \sum r_i \otimes f_i + \sum s_j \otimes h_j$, for $x \in G$, we have $x.f = \sum r_i(x^{-1}) f_i + \sum s_j(x^{-1}) h_j$. Since this element belongs to F, the functions s_j must vanish identically on G, hence must be 0. We then have $\varphi^* F \subset C[G] \otimes_C F$. □

Theorem 3.4.3. *Let G be an affine algebraic group. Then G is isomorphic to a closed subgroup of some* $\mathrm{GL}(n, C)$.

Proof. Choose generators $f_1, \ldots, f_n$ for the coordinate algebra $C[G]$. By applying Lemma 3.4.2 a), we can assume that the f_i are a C-basis of a C-vector space F which is G-stable when considering the action of G by right translations. If $\varphi : G \times G \to G$ is given by $(x, y) \mapsto yx$, by Lemma 3.4.2 b), we can write $\varphi^* f_i = \sum_j m_{ij} \otimes f_j$, where $m_{ij} \in C[G]$. Then $\rho_x(f_i)(y) = f_i(yx) = \sum_j m_{ij}(x) f_j(y)$; whence $\rho_x(f_i) = \sum_j m_{ij}(x) f_j$. In other words, the matrix of $\rho_x | F$ in the basis $\{f_i\}$ is $(m_{ij}(x))$. This shows that the map $\psi : G \to \mathrm{GL}(n, C)$ defined by $x \mapsto (m_{ij}(x))$ is a morphism of algebraic groups.

Notice that $f_i(x) = f_i(ex) = \sum m_{ij}(x) f_j(e)$, i.e. $f_i = \sum f_j(e) m_{ij}$. This shows that the m_{ij} also generate $C[G]$; in particular, ψ is injective. Moreover the image group $G' = \psi(G)$ is closed in $\mathrm{GL}(n, C)$ by Proposition 3.3.4 b). To complete the proof we therefore need to show only that $\psi : G \to G'$ is an isomorphism of varieties. But the restriction to G' of the coordinate

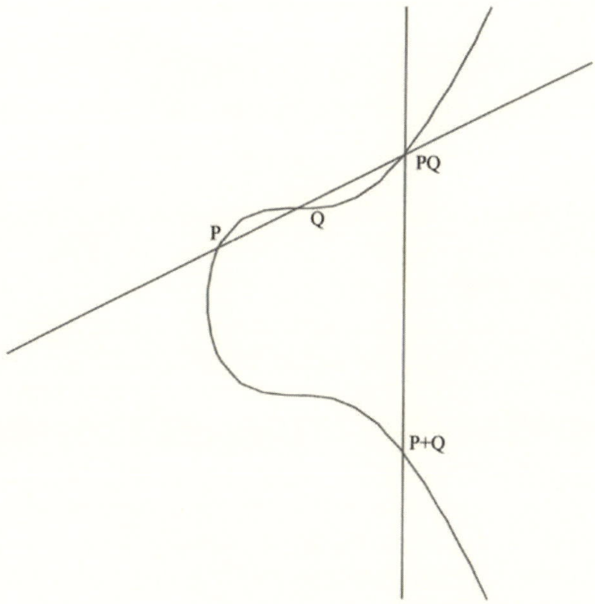

Figure 1. Sum of two points.

functions X_{ij} are sent by ψ^* to the respective m_{ij}, which were just shown to generate $C[G]$. So ψ^* is surjective and thus identifies $C[G']$ with $C[G]$. $\square$

We now give an example of an algebraic group which is not affine, hence not a linear algebraic group.

Example 3.4.4. We consider the curve E in the projective plane $\mathbb{P}^2_C$ given by the equation

$$Y^2Z = X^3 + aXZ^2 + bZ^3,$$

with $4a^3 + 27b^2 \neq 0$. We have seen (Exercise 9 in chapter 2) that E is nonsingular, has a unique point $\mathbf{0}$ at infinity, namely $\mathbf{0} = (0 : 1 : 0)$, and that the tangent line to E at $\mathbf{0}$ is the line at infinity $Z = 0$. Moreover, the intersection of E with $Z = 0$ has the point $\mathbf{0}$ as a triple solution. The set $E(C)$ of the points of E with coordinates in C is then the set of points with coordinates in C of the affine plane curve with equation

$$Y^2 = X^3 + aX + b$$

plus the point $\mathbf{0}$. A plane nonsingular cubic is called an *elliptic curve.* We now define a sum on $E(C)$. Let us note that the intersection of E with a line

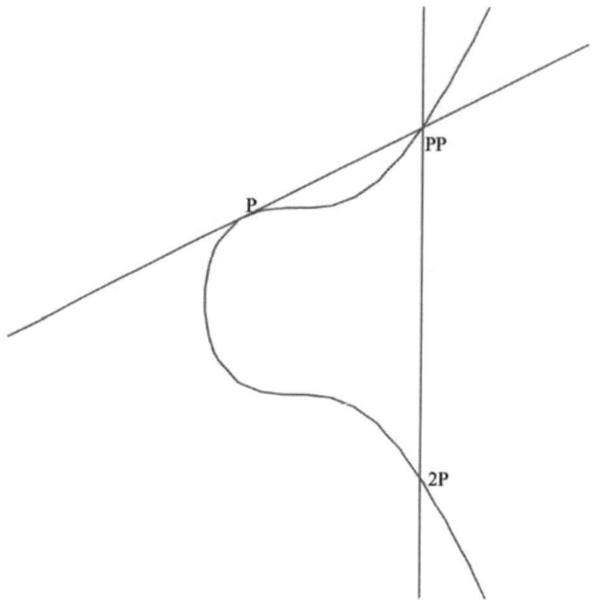

Figure 2. Doubling of a point.

l defined over C consists in three points, distinct or not, with coordinates in C. If P, Q are two distinct points on E (resp. if P is a point on E), denote by l the line joining P and Q (resp. the tangent to E at P). Let PQ (resp. PP) be the third point in the intersection $l \cap E$. (If l is the tangent at P, we count P twice.) Let $P+Q$ (resp. $2P$) be the third point in the intersection of E with the line joining PQ (resp. PP) and $\mathbf{0}$. (See Figures 1 and 2.) It is clear that this sum is commutative, has $\mathbf{0}$ as identity element and that for $P = (x_P, y_P) \in E$, $-P = (x_P, -y_P)$. Let us now write this sum in coordinates. We distinguish three cases.

(1) If $P = (x_P, y_P) \neq Q = (x_Q, y_Q)$ and $x_P \neq x_Q$, the equation of the line l is $Y = \alpha X + \beta$, with $\alpha = (y_Q - y_P)/(x_Q - x_P), \beta = (y_P x_Q - y_Q x_P)/(x_Q - x_P)$. Substituting in the equation of E, we obtain a monic polynomial in X of degree 3, in which the coefficient of X^2 is $-\alpha^2$, so the coordinates of $P+Q$ are

$$x_{P+Q} = \alpha^2 - x_P - x_Q, y_{P+Q} = -\alpha x_{P+Q} - \beta.$$

(2) If $P \neq Q$ and $x_P = x_Q$, $P + Q = \mathbf{0}$.

(3) The equation of the tangent line to E at the point $P = (x_P, y_P)$ is $Y = \alpha X + \beta$, with $\alpha = (3x_P^2 + a)/2y_P, \beta = (-x_P^3 + ax_P + 2b)/2y_P$. Again substituting in the equation of E, we obtain a monic polynomial in X of degree 3, in which the coefficient of X^2 is $-\alpha^2$, so the coordinates of $2P$ are

$$x_{2P} = \alpha^2 - 2x_P, y_{2P} = -\alpha x_{2P} - \beta.$$

The more delicate point is proving that the defined law is associative. One tedious form to do it is by direct computation using the formulae above. A geometric proof based on the fact that a pencil of cubic curves is determined by eight points can be found in [**Fu**] or [**Hus**]. A third proof using Riemann-Roch theorem can be found in [**Si**].

We then have on the projective curve E a group structure and the formulae obtained above show that the maps giving the sum and the opposite element are morphisms of varieties; hence E is an algebraic group.

3.5. Homogeneous spaces

Let G be an algebraic group. A *homogeneous space* for G is a G-variety V on which G acts transitively. An example of homogeneous space for G is $V = G$, with the action given by left or right translations introduced in Section 3.4.

Lemma 3.5.1. *Let V be a G-variety.*

a) For $v \in V$, the orbit $G.v$ is open in its closure.

b) There exist closed orbits.

Proof. By applying Proposition 2.2.13 to the morphism $G \to V$, $x \mapsto x.v$, we obtain that $G.v$ contains a nonempty open subset U of its closure. Since $G.v$ is the union of the open sets $x.U, x \in G$, assertion a) follows. It implies that for $v \in V$, the set $S_v = \overline{G.v} \setminus G.v$ is closed. It is also G-stable, hence a union of orbits. As the descending chain condition on closed sets is satisfied, there is a minimal set S_v. By a), it must be empty. Hence the orbit $G.v$ is closed, proving b). □

Lemma 3.5.2. *Let G be an algebraic group and G^0 its identity component. Let V be a homogeneous space for G.*

a) Each irreducible component of V is a homogeneous space for G^0.

b) The components of V are open and closed and V is their disjoint union.

Proof. Let V' be the orbit of G^0 in V. Since G acts transitively on V, it follows from Proposition 3.2.1 that V is the disjoint union of finitely many translates $x.V'$. Each of them is a G^0-orbit and is irreducible. It follows from Lemma 3.5.1 that all G^0-orbits are closed. Now the xV' are the irreducible components of V and each of them is a G_0-orbit, so a) is proved. As V is the disjoint union of the closed sets xV' and there is a finite number of them, b) follows. □

Proposition 3.5.3. *Let G be an algebraic group and let $\varphi : V \to W$ be an equivariant morphism of homogeneous spaces for G. Put $r = \dim V - \dim W$.*

a) For any variety Z the morphism $(\varphi, Id) : V \times Z \to W \times Z$ is open.

b) If W' is an irreducible closed subvariety of W and V' an irreducible component of $\varphi^{-1}W'$, then $\dim V' = \dim W' + r$. In particular, if $y \in W$, then all irreducible components of $\varphi^{-1}y$ have dimension r.

Proof. Using Lemma 3.5.2, we reduce the proof to the case when G is connected and V, W are irreducible. Then φ is surjective, hence dominant. Let $U \subset V$ be an open subset with the properties of Proposition 2.2.24 and Corollary 2.2.25. Then all translates $x.U, x \in G$, enjoy the same properties. Since these cover V, we have a) and b). □

A homogeneous space V for G is a *principal homogeneous space* or *G-torsor* if the morphism

$$\begin{array}{ccc} G \times V & \to & V \times V \\ (x, v) & \mapsto & (x.v, v) \end{array}$$

is an isomorphism. The action of G on V is then transitive and the stabilizer subgroup of any element in V is trivial. Clearly, $V = G$ with the action given by left translations is a G-torsor. It is called the trivial G-torsor. More generally, we say that a G-torsor is trivial if it is G-isomorphic to G. In particular, G with the action given by right translations is a trivial G-torsor.

3.6. Characters and semi-invariants

Definition 3.6.1. A *character* of an algebraic group G is a morphism of algebraic groups $G \to \mathbb{G}_m$.

For example, the determinant defines a character of $\mathrm{GL}(n, C)$. If χ_1, χ_2 are characters of an algebraic group G, so is their product defined by $(\chi_1\chi_2)(x) = \chi_1(x)\chi_2(x)$. This product gives the set $X(G)$ of all characters of G the structure of a commutative group. The identity element is the character χ_0 such that $\chi_0(x) = 1$ for all $x \in G$.

Example 3.6.2. A morphism $\chi : \mathbb{G}_a \to \mathbb{G}_m$ would be given by a polynomial $\chi(x)$ satisfying $\chi(x + y) = \chi(x)\chi(y)$. We then obtain $X(\mathbb{G}_a) = 1$.

Example 3.6.3. Given a character χ of $\mathrm{SL}(n, C)$, $n \geq 2$, by composition with the morphism $\mathbb{G}_a \to \mathrm{SL}(n, C)$, $x \mapsto I + xe_{ij}$, where we denote by $e_{ij}, i \neq j$, the matrix with entry 1 in the position (i, j) and 0's elsewhere, we obtain a character of $\mathbb{G}_a$. As the subgroups $U_{ij} = \{I + xe_{ij} : x \in C\}$ generate $\mathrm{SL}(n, C)$, we obtain $X(\mathrm{SL}(n, C)) = 1$.

Example 3.6.4. A character of $\mathbb{G}_m$ is defined by $x \mapsto x^n$, for some $n \in \mathbb{Z}$; hence $X(\mathbb{G}_m) \simeq \mathbb{Z}$. As $\mathrm{D}(n, C) \simeq \mathbb{G}_m \times \cdots \times \mathbb{G}_m$, we obtain $X(\mathrm{D}(n, C)) \simeq \mathbb{Z} \times \cdots \times \mathbb{Z}$.

If G is a closed subgroup of $\mathrm{GL}(V)$, for each $\chi \in X(G)$, we define $V_\chi = \{v \in V : x.v = \chi(x)v\,,\ \text{for all } x \in G\}$. Evidently V_χ is a G-stable subspace of V. Any nonzero element of V_χ is called a *semi-invariant* of G of *weight* χ. Conversely if v is any nonzero vector which spans a G-stable line in V, then it is clear that $x.v = \chi(x)v$ defines a character χ of G.

More generally, if $\varphi : G \to \mathrm{GL}(V)$ is a rational representation, i.e. a rational map which is also a group homomorphism, then the semi-invariants of G are by definition those of $\varphi(G)$.

Lemma 3.6.5. *Let $\varphi : G \to \mathrm{GL}(V)$ be a rational representation. Then the subspaces $V_\chi, \chi \in X(G)$, are in direct sum; in particular, only finitely many of them are nonzero.*

Proof. Otherwise, we could choose a minimal $n \geq 2$ and nonzero vectors $v_i \in V_{\chi_i}$, for distinct $\chi_i, 1 \leq i \leq n$, such that $v_1 + \cdots + v_n = 0$. Since the χ_i are distinct, $\chi_1(x) \neq \chi_2(x)$ for some $x \in G$. But $0 = \varphi(x)(\sum v_i) = \sum \chi_i(x)v_i$, so $\sum \chi_1(x)^{-1}\chi_i(x)v_i = 0$. The coefficient of v_2 is different from 1; so we can subtract this equation from the equation $\sum v_i = 0$ to obtain a nontrivial dependence involving $\leq n - 1$ characters, contradicting the choice of n. □

Lemma 3.6.6. *Let $\varphi : G \to \mathrm{GL}(V)$ be a rational representation. Let H be a closed normal subgroup of G. Then each element of $\varphi(G)$ permutes the spaces V_χ for $\chi \in X(H)$.*

Proof. We can assume that $G \subset \mathrm{GL}(V)$. If $x \in G, y \in H, v \in V_\chi$, then $y.(x.v) = (yx).v = x(x^{-1}yx).v = x.(\chi(x^{-1}yx).v) = \chi(x^{-1}yx)x.v$ and the function $y \mapsto \chi(x^{-1}yx)$ is clearly a character χ' of H, so x maps V_χ into $V_{\chi'}$. □

3.7. Quotients

The aim of this section is to prove that if G is a linear algebraic group and H a closed normal subgroup of G, then the quotient G/H has the natural structure of a linear algebraic group, with coordinate ring $C[G/H] \simeq C[G]^H$.

If V is a finite dimensional C-vector space, then $\mathrm{GL}(V)$ acts on exterior powers of V by $x.(v_1 \wedge \cdots \wedge v_k) = x.v_1 \wedge \cdots \wedge x.v_k$. If M is a d-dimensional subspace of V, it is especially useful to look at the action on $L = \wedge^d M$, which is a 1-dimensional subspace of $\wedge^d V$.

Lemma 3.7.1. *For $x \in \mathrm{GL}(V)$, M a d-dimensional subspace of V, $L = \wedge^d M$, we have $xL = L$ if and only if $xM = M$.*

Proof. The "if" part is clear. For the other implication, we can choose a basis $v_1, \dots, v_n$ in V such that $v_1, \dots, v_l$ is a basis of $M \cap xM$, $v_1, \dots, v_d$ is a basis of M, and $v_1, \dots, v_l, v_{d+1}, \dots, v_{2d-l}$ is a basis of xM. By hypothesis $x(v_1 \wedge \cdots \wedge v_d)$ is a multiple of $v_1 \wedge \cdots \wedge v_d$ but, on the other hand, it is a multiple of $v_1 \wedge \cdots \wedge v_l \wedge v_{d+1} \wedge \dots, v_{2d-l}$ forcing $l = d$. □

Proposition 3.7.2. *Let G be a linear algebraic group, H a closed subgroup of G. Then there is a rational representation $\varphi : G \to \mathrm{GL}(V)$ and a 1-dimensional subspace L of V such that $H = \{x \in G : \varphi(x)L = L\}$.*

Proof. Let I be the ideal in $C[G]$ vanishing on H. It is a finitely generated ideal. By Lemma 3.4.2, there exists a finite dimensional subspace W of $C[G]$, stable under all $\rho_x, x \in G$, which contains a given finite generating set of I. Set $M = W \cap I$, so M generates I. Notice that M is stable under all $\rho_x, x \in H$, since by Lemma 3.4.1, $H = \{g \in G : \rho_g I = I\}$. We claim that $H = \{x \in G : \rho_x M = M\}$. Assume that we have $\rho_g M = M$. As M generates I, we have $\rho_x I = I$; hence $x \in H$.

Now take $V = \wedge^d W, L = \wedge^d M$, for $d = \dim M$. By Lemma 3.7.1, we have the desired characterization of H. □

Theorem 3.7.3. *Let G be a linear algebraic group, H a closed normal subgroup of G. Then there is a rational representation $\psi : G \to \mathrm{GL}(W)$ such that $H = \mathrm{Ker}\,\psi$.*

Proof. By Proposition 3.7.2, there exists a morphism $\varphi : G \to \mathrm{GL}(V)$ and a line L such that $H = \{x \in G : \varphi(x)L = L\}$. Since each element in H acts on L by scalar multiplication, this action has an associated character $\chi_0 : H \to \mathbb{G}_m$. Consider the sum in V of all nonzero V_χ for all characters χ of H. By Lemma 3.6.5, this sum is direct and of course includes L. Moreover, by Lemma 3.6.6, $\varphi(G)$ permutes the various V_χ. So we can assume that V itself is the sum of the V_χ.

Now let W be the subspace of $\mathrm{End}\, V$ consisting of those endomorphisms which stabilize each V_χ, $\chi \in X(H)$. There is a natural isomorphism $W \simeq \oplus \mathrm{End}\, V_\chi$. Now $\mathrm{GL}(V)$ acts on $\mathrm{End}\, V$ by conjugation. Notice that the subgroup $\varphi(G)$ stabilizes W, since $\varphi(G)$ permutes the V_χ and W stabilizes each of them. We then obtain a group morphism $\psi : G \to \mathrm{GL}(W)$ given by $\psi(x)(y) = \varphi(x)_{|W}\, y\, \varphi(x)^{-1}_{|W}$; so ψ is a rational representation. Let us now check $H = \operatorname{Ker}\psi$. If $x \in H$, then $\varphi(x)$ acts as a scalar on each V_χ, so conjugating by $\varphi(x)$ has no effect on W; hence $x \in \operatorname{Ker}\psi$. Conversely, let $x \in G$, $\psi(x) = I$. This means that $\varphi(x)$ stabilizes each V_χ and commutes with $\operatorname{End} V_\chi$. But the center of $\operatorname{End} V_\chi$ is the set of scalars, so $\varphi(x)$ acts on each V_χ as a scalar. In particular, $\varphi(x)$ stabilizes $L \subset V_{\chi_0}$, forcing $x \in H$. □

Corollary 3.7.4. *The quotient G/H can be given a structure of linear algebraic group endowed with an epimorphism $\pi : G \to G/H$.*

Proof. We consider the representation $\psi : G \to \mathrm{GL}(W)$ with kernel H given by Theorem 3.7.3 and its image $Y = \operatorname{Im}\psi$. By Theorem 2.2.21, Y is a constructible set and, as it is a subgroup of $\mathrm{GL}(W)$, by Proposition 3.3.2, it is a closed subgroup of $\mathrm{GL}(W)$. We have a group isomorphism $G/H \simeq Y$; hence we can translate the linear algebraic group structure of Y to G/H. Moreover ψ induces an epimorphism of algebraic groups $\pi : G \to G/H$. □

Definition 3.7.5. Let G be an algebraic group, H a closed subgroup of G. A *Chevalley quotient* of G by H is a variety X together with a surjective morphism $\pi : G \to X$ such that the fibers of π are exactly the cosets of H in G.

In Corollary 3.7.4, we have established that there exists a Chevalley quotient of a linear algebraic group G by a closed normal subgroup H. However, it is not clear if Chevalley quotients are unique up to isomorphism nor if they satisfy the usual universal property of quotients. These properties characterize categorical quotients which we define next.

Definition 3.7.6. Let G be an algebraic group, H a closed subgroup of G. A *categorical quotient* of G by H is a variety X together with an epimorphism $\pi : G \to X$ that is constant on all cosets of H in G with the following universal property: given any other variety Y and a morphism $\varphi : G \to Y$ that is constant on all cosets of H in G there is a unique morphism $\overline{\varphi} : X \to Y$ such that $\varphi = \overline{\varphi} \circ \pi$.

It is clear that categorical quotients are unique up to unique isomorphism. Our aim is to prove that Chevalley quotients are categorical quotients. We then will have a quotient of G by H defined uniquely up to isomorphism and satisfying the universal property.

Theorem 3.7.7. *Chevalley quotients are categorical quotients.*

Proof. We first construct a categorical quotient in the category of geometric spaces. Define G/H to be the set of cosets of H in G. Let $\pi : G \to G/H$

be the map defined by $x \mapsto xH$. Give G/H the structure of a topological space by defining $U \subset G/H$ to be open if and only if $\pi^{-1}(U)$ is open in G. Next define a sheaf $\mathcal{O} = \mathcal{O}_{G/H}$ of C-valued functions on G/H as follows: if $U \subset G/H$ is open, then $\mathcal{O}(U)$ is the ring of functions f on U such that $f \circ \pi$ is regular on $\pi^{-1}(U)$. (This defines indeed a sheaf of functions.) In order to check the universal property, let $\psi : G \to Y$ be a morphism of geometric spaces constant on the cosets of H in G. We get the induced map of sets $\overline{\psi} : G/H \to Y$, $xH \mapsto \psi(x)$, clearly satisfying $\psi = \overline{\psi} \circ \pi$. We prove that $\overline{\psi}$ is a morphism of geometric spaces. To check continuity, take an open subset $V \subset Y$ and note that $U := \overline{\psi}^{-1}(V)$ is open in G/H, by the definition of the topology in G/H and the continuity of ψ. Finally, for $f \in \mathcal{O}_Y(V)$, $\overline{\psi}^*(f) \in \mathcal{O}_{G/H}$, because $\pi^*(\overline{\psi}^*(f)) \in \mathcal{O}_G(\psi^{-1}(V))$.

Now we take $(G/H, \pi)$ as above and let (X, ψ) be a Chevalley quotient. Using the universal property established above, we get a unique G-equivariant morphism $\overline{\psi} : G/H \to X$ such that $\psi = \overline{\psi} \circ \pi$. We will prove that $\overline{\psi}$ is an isomorphism of geometric spaces, which will imply that G/H is a variety and that X is a categorical quotient.

By Lemma 3.5.2, we can assume that G is a connected algebraic group. First of all, it is clear that $\overline{\psi}$ is a continuous bijection. If $U \subset G/H$ is open, then $\overline{\psi}(U) = \psi(\pi^{-1}(U))$ and by Proposition 3.5.3 a), it follows that $\overline{\psi}(U)$ is open, which implies that $\overline{\psi}$ is a homeomorphism.

In order to prove that $\overline{\psi}$ is an isomorphism, the following has to be established: If U is an open set in X, the homomorphism of C-algebras $\mathcal{O}_X(U) \to \mathcal{O}_{G/H}(\overline{\psi}^{-1}(U))$ defined by $\overline{\psi}^*$ is an isomorphism. By definition of $\mathcal{O}_{G/H}$ this means that, for any regular function f on $V = \psi^{-1}(U)$ such that $f(gh) = f(g)$, $\forall g \in V, h \in H$, there is a unique regular function F on U such that $F(\psi(g)) = f(g)$. Let $\Gamma = \{(g, f(g)) : g \in V\} \subset V \times \mathbb{A}^1$ be the graph of f and put $\Gamma' = (\psi, Id)(\Gamma)$, so $\Gamma' \subset U \times \mathbb{A}^1$. Since Γ is closed in $V \times \mathbb{A}^1$, Proposition 3.5.3 a) shows that $(\psi, Id)(V \times \mathbb{A}^1 \setminus \Gamma) = U \times \mathbb{A}^1 \setminus \Gamma'$ is open in $U \times \mathbb{A}^1$. Hence Γ' is closed in $U \times \mathbb{A}^1$. Let $\lambda : \Gamma' \to U$ be the morphism induced by the projection on the first component. It follows from the definition that λ is bijective and birational. By Zariski's Main Theorem 2.2.32, λ is an isomorphism. This implies that there exists a regular function F on U such that $\Gamma' = \{(u, F(u)) : u \in U\}$, which is what we wanted to prove. This finishes the proof of the theorem. □

We recall that the action of G on itself by translation on the left gives an action of G on its coordinate ring $C[G]$ defined by $\lambda_x(f)(y) = f(x^{-1}y)$ for $f \in C[G], x, y \in G$. (See (3.1).)

Proposition 3.7.8. *Let G be a linear algebraic group, H a closed normal subgroup of G. We have $C[G/H] \simeq C[G]^H$.*

Proof. We consider the epimorphism π given by Corollary 3.7.4. If $f \in C[G/H]$, then $\widetilde{f} = f \circ \pi \in C[G]$. Moreover, for $x \in H, y \in G$, we have $\lambda_x(\widetilde{f})(y) = \widetilde{f}(x^{-1}y) = (f \circ \pi)(x^{-1}y) = f(\pi(x^{-1}y)) = f(\pi(y)) = \widetilde{f}(y)$, so $\lambda_x(\widetilde{f}) = \widetilde{f}$ and $\widetilde{f} \in C[G]^H$.

If $f \in C[G]^H$, then f is a morphism $G \to \mathbb{A}^1$ which is constant on the cosets of H in G. Hence, by the universal property of the quotient G/H established in Theorem 3.7.7, there exists $F \in C[G/H]$ such that $f = F \circ \pi$. $\square$

Exercises

(1) Determine the dimension of each of the linear algebraic groups $\mathrm{GL}(n,C), \mathrm{SL}(n,C), \mathrm{T}(n,C), \mathrm{U}(n,C)$, and $\mathrm{D}(n,C)$.

(2) Determine the dimension of the orthogonal group

$$\mathrm{O}(n,C) := \{A \in \mathrm{GL}(n,C) : AA^T = Id\}$$

and of the special orthogonal group

$$\mathrm{SO}(n,C) := \{A \in \mathrm{O}(n,C) : \det A = 1\}.$$

(3) Let G be an algebraic group, H a subgroup of G. Prove that
a) If H is commutative, so is $\overline{H}$.
b) If H is normal in G, so is $\overline{H}$.

(4) Let G_1, G_2 be algebraic groups, $M_1, P_1 \subset G_1, M_2 \subset G_2$ and $\varphi : G_1 \to G_2$ a morphism of algebraic groups. We denote by $\mathrm{GC}(M)$ the group closure of a subset M of an algebraic group G. We recall that for a subset S of a group G we define the *centralizer* $C_G(S)$ *of* S *in* G as

$$C_G(S) = \{x \in G : xsx^{-1} = s, \ \forall s \in S\}.$$

If a subgroup H of G is contained in $C_G(S)$, we say that H *centralizes* S.
Prove that
a) If M_1 is a dense subset of P_1, then $GC(M_1) = GC(P_1)$.
b) If M_1 normalizes (resp. centralizes) P_1, then $GC(M_1)$ normalizes (resp. centralizes) $GC(P_1)$.
c) $GC(M_1 \times M_2) = GC(M_1) \times GC(M_2)$.
d) $\varphi(GC(M_1)) = GC(\varphi(M_1))$.

(5) Let G be an affine algebraic group, let $\mu : G \times G \to G$, $\iota : G \to G$ be the morphisms giving the group structure, and let e be the morphism from the trivial group into G, p the constant morphism from G in G sending all elements to e.
a) Express the group axioms for G in terms of commutative diagrams of morphisms of affine varieties involving the morphisms μ, ι, e, p and the identity automorphism of G.
b) Write down the commutative diagrams of morphisms of C-algebras obtained from the diagrams in a) by passing from the affine varieties morphisms to the associated morphisms between the corresponding coordinate rings. We obtain then that $A = C[G]$ is endowed with a structure given by morphisms $\mu^* : A \to A \otimes A$, $\iota^* : A \to A$, $e^* : A \to C$, making commutative the diagrams obtained. Give the

expressions of the images of a function $f \in C[G]$ by the preceding morphisms.

A C-algebra A with algebra morphisms μ^*, e^*, ι^* making commutative the above diagrams is called a *Hopf algebra*, the morphism μ^* is called *coproduct*, e^* *counit* and ι^* *coinverse.*

c) Write down explicitly the morphisms μ^*, ι^* and e^* corresponding to the additive group, the multiplicative group, and the general linear group.

(6) Let X be an algebraic variety.

a) If G is an algebraic group, prove that the set $\mathrm{Hom}(X, G)$ of algebraic varieties morphisms from X to G has a natural group structure.

b) Show that $\mathrm{Hom}(X, \mathbb{G}_a)$ is isomorphic to $\mathcal{O}(X)$ as a group under addition.

c) Show that $\mathrm{Hom}(X, \mathbb{G}_m)$ is isomorphic to the group of units in $\mathcal{O}(X)$ under multiplication.

(7) *Burnside Theorem.* Let E be a finite dimensional C-vector space and A a subalgebra of $\mathrm{End}(E)$. Prove that if the only A-stable subspaces of E are 0 and E, then $A = \mathrm{End}(E)$.

Hint: Prove that the minimal rank of the non-zero elements in A is 1 and that A contains all endomorphisms of rank 1.

(8) Let E be a finite dimensional C-vector space. An element x in $\mathrm{GL}(E)$ is called unipotent if $(x - I)^n = 0$ for some integer $n > 0$, where I denotes identity. A subgroup G of $\mathrm{GL}(E)$ is unipotent if all its elements are unipotent. If G is an unipotent subgroup of $\mathrm{GL}(E)$, prove that there exists a nonzero vector in E which is fixed by all elements in G. Deduce that a unipotent subgroup of $\mathrm{GL}(n, C)$ is conjugate to a subgroup of $U(n, C)$.

(9) Let G be an affine algebraic group. Prove that if G is connected and $\dim G = 1$, then G is commutative.

(10) Prove that a closed subgroup H of the upper triangular unipotent group $U(n, C)$ has no nontrivial characters.

Hint: A nontrivial character of H would give a nontrivial character of the additive group $\mathbb{G}_a(C)$.

(11) Let H_1, H_2 be closed normal subgroups of an algebraic group G such that $H_1 \subset H_2$. Let $\pi : G \to G/H_1$ be the canonical surjection.

a) Prove that $\pi(H_2)$ is a closed normal subgroup of G/H_1, isomorphic to the algebraic group H_2/H_1.

b) Prove that the algebraic groups G/H_2 and $(G/H_1)/\pi(H_2)$ are isomorphic.

Chapter 4

Lie Algebras and Algebraic Groups

Lie algebras were introduced to study infinitesimal transformations. Our aim here is to define the Lie algebra associated to a linear algebraic group. We point out that some properties of an algebraic group are more easily read in its Lie algebra. An important property of an algebraic group is its solvability. We shall see that the solvability by quadratures of a homogeneous linear differential equation is characterized by the solvability of the identity component of its differential Galois group. In this chapter, we present the Lie-Kolchin theorem, which states that the connected solvable linear algebraic groups are exactly the triangularizable ones and will be essential in proving the characterization of solvability by quadratures by means of the differential Galois group. We shall establish that the solvability of a connected linear algebraic group is equivalent to the solvability of its Lie algebra. We conclude the chapter with the classification of the subgroups of the special linear group $\mathrm{SL}(2,\mathbb{C})$ which will be used in Kovacic's algorithm.

4.1. Lie algebras

Definition 4.1.1. A *Lie algebra* over a field C is a C-vector space $\mathfrak{g}$ together with a binary operation

$$[\cdot,\cdot]:\mathfrak{g}\times\mathfrak{g}\to\mathfrak{g}$$

called the *Lie bracket* , which satisfies the following axioms:

(1) Bilinearity:

$$[ax + by, z] = a[x, z] + b[y, z], \quad [z, ax + by] = a[z, x] + b[z, y],$$

for all scalars $a, b \in C$ and all elements $x, y, z \in \mathfrak{g}$.

(2) Alternating:

$$[x, x] = 0$$

for all $x \in \mathfrak{g}$. This implies anticommutativity, i.e.

$$[x, y] = -[y, x]$$

for all elements $x, y \in \mathfrak{g}$.

(3) The Jacobi identity:

$$[x, [y, z]] + [y, [z, x]] + [z, [x, y]] = 0$$

for all $x, y, z \in \mathfrak{g}$.

Example 4.1.2. Any C-vector space $\mathfrak{g}$ can be given a Lie algebra structure with the trivial Lie bracket $[x, y] = 0$, for all $x, y \in \mathfrak{g}$. Such a Lie algebra is called *abelian* or *commutative.*

Let $\mathfrak{g}$ be a Lie algebra. A subspace $\mathfrak{h}$ of $\mathfrak{g}$ is a *Lie subalgebra* of $\mathfrak{g}$ (resp. an *ideal* of $\mathfrak{g}$) if $[x, y] \in \mathfrak{h}$ for all $x, y \in \mathfrak{h}$ (resp. for all $x \in \mathfrak{g}, y \in \mathfrak{h}$). If $\mathfrak{h}$ is an ideal of $\mathfrak{g}$, the quotient $\mathfrak{g}/\mathfrak{h}$ inherits a natural structure of Lie algebra given by $[x + \mathfrak{h}, y + \mathfrak{h}] = [x, y] + \mathfrak{h}$.

Example 4.1.3. To any associative C-algebra A, one can associate a Lie algebra $L(A)$. As a C-vector space, $L(A)$ is the same as A. The Lie bracket is defined from the product in A by $[x, y] := xy - yx$. The associativity of the product implies the Jacobi identity for the bracket. In particular, the algebra of $n \times n$ matrices over C gives rise to a Lie algebra called the *general linear Lie algebra* and denoted $\mathfrak{gl}(n, C)$. Equivalently, if V is a finite dimensional C-vector space, the algebra of endomorphisms of V gives rise to the Lie algebra $\mathfrak{gl}(V)$. More generally, a C-vector subspace of an associative C-algebra which is closed under the Lie bracket $[x, y] := xy - yx$ is a Lie algebra. The following subspaces of $\mathfrak{gl}(n, C)$ are clearly Lie subalgebras:

- the subspace of the matrices in $\mathfrak{gl}(n, C)$ whose trace is zero, denoted by $\mathfrak{sl}(n, C)$;
- the subspace of upper triangular matrices in $\mathfrak{gl}(n, C)$, denoted by $\mathfrak{t}(n, C)$;
- the subspace of strictly upper triangular matrices in $\mathfrak{gl}(n, C)$, denoted by $\mathfrak{n}(n, C)$;
- the subspace of diagonal matrices in $\mathfrak{gl}(n, C)$, denoted by $\mathfrak{d}(n, C)$.

Moreover $\mathfrak{sl}(n, C)$ is an ideal of $\mathfrak{gl}(n, C)$. The notation used is justified in Example 4.2.8 below.

If $\mathfrak{h}, \mathfrak{h}'$ are ideals of $\mathfrak{g}$, then, using the Jacobi identity, we see that $[\mathfrak{h}, \mathfrak{h}'] := \{[x, x'] : x \in \mathfrak{h}, x' \in \mathfrak{h}'\}$ is also an ideal of $\mathfrak{g}$. In particular, $[\mathfrak{g}, \mathfrak{g}]$ is an ideal of $\mathfrak{g}$.

If $\mathfrak{p}, \mathfrak{q}$ are subsets of $\mathfrak{g}$, we define the *centralizer* of $\mathfrak{p}$ in $\mathfrak{q}$ by

$$\mathfrak{c}_{\mathfrak{q}}\mathfrak{p} := \{x \in \mathfrak{q} : [x, y] = 0\,,\ \forall y \in \mathfrak{p}\}.$$

In particular, $\mathfrak{z}(\mathfrak{g}) = \mathfrak{c}_{\mathfrak{g}}\mathfrak{g}$ is called the *center* of $\mathfrak{g}$. We have $\mathfrak{g}$ abelian if and only if $\mathfrak{z}(\mathfrak{g}) = \mathfrak{g}$.

The *normalizer* of $\mathfrak{p}$ in $\mathfrak{q}$ is defined by

$$\mathfrak{n}_{\mathfrak{q}}\mathfrak{p} := \{x \in \mathfrak{q} : [x, y] \in \mathfrak{p}\,,\ \forall y \in \mathfrak{p}\}.$$

If $\mathfrak{p}$ is a Lie subalgebra, then so is $\mathfrak{n}_{\mathfrak{g}}\mathfrak{p}$, and $\mathfrak{p}$ is an ideal of $\mathfrak{n}_{\mathfrak{g}}\mathfrak{p}$.

If $\mathfrak{g}, \mathfrak{g}'$ are Lie algebras, a linear map $\varphi : \mathfrak{g} \to \mathfrak{g}'$ is a *morphism of Lie algebras* if $\varphi([x, y]) = [\varphi x, \varphi y]$, for all $x, y \in \mathfrak{g}$. A linear map $d : \mathfrak{g} \to \mathfrak{g}$ is called a *derivation* of $\mathfrak{g}$ if for all $x, y \in \mathfrak{g}$

$$d([x, y]) = [d(x), y] + [x, d(y)].$$

We denote by $\operatorname{Der}\mathfrak{g}$ the C-vector space of derivations of $\mathfrak{g}$. It has a natural structure of Lie algebra with the Lie bracket defined by

$$[d, d'] = d \circ d' - d' \circ d, \text{ for } d, d' \in \operatorname{Der}\mathfrak{g}.$$

Let $x \in \mathfrak{g}$. The Jacobi identity and the anticommutativity of the Lie bracket imply that the linear map $y \mapsto [x, y]$, denoted by $\operatorname{ad}_{\mathfrak{g}} x$ or $\operatorname{ad} x$ is a derivation of $\mathfrak{g}$ called an *inner derivation* of $\mathfrak{g}$. By the bilinearity of the Lie bracket, the map $\mathfrak{g} \to \operatorname{Der}\mathfrak{g}, x \mapsto [x, \cdot]$ is linear. It is called the *adjoint representation* of $\mathfrak{g}$.

An ideal of $\mathfrak{g}$ is said to be *characteristic* if it is invariant under all the derivations of $\mathfrak{g}$. We define by induction two decreasing chains of characteristic ideals of $\mathfrak{g}$.

$$\mathcal{C}^1(\mathfrak{g}) = \mathfrak{g}, \quad \mathcal{C}^2(\mathfrak{g}) = [\mathfrak{g}, \mathfrak{g}], \dots, \quad \mathcal{C}^{i+1}(\mathfrak{g}) = [\mathfrak{g}, \mathcal{C}^i(\mathfrak{g})], \dots$$

$$\mathcal{D}^0(\mathfrak{g}) = \mathfrak{g}, \quad \mathcal{D}^1(\mathfrak{g}) = [\mathfrak{g}, \mathfrak{g}], \dots, \quad \mathcal{D}^{i+1}(\mathfrak{g}) = [\mathcal{D}^i(\mathfrak{g}), \mathcal{D}^i(\mathfrak{g})], \dots$$

We call $(\mathcal{C}^i(\mathfrak{g}))_{i\geq 1}$ the *central descending series* of $\mathfrak{g}$ and $(\mathcal{D}^i(\mathfrak{g}))_{i\geq 0}$ the *derived series* of $\mathfrak{g}$. In particular $\mathcal{D}(\mathfrak{g}) := \mathcal{D}^1(\mathfrak{g})$ is the *derived ideal* of $\mathfrak{g}$.

If $\varphi : \mathfrak{g} \to \mathfrak{g}'$ is a Lie algebra morphism, then $\varphi(\mathcal{C}^i(\mathfrak{g})) \subset \mathcal{C}^i(\mathfrak{g}')$ and $\varphi(\mathcal{D}^i(\mathfrak{g})) \subset \mathcal{D}^i(\mathfrak{g}')$ for all i. Moreover, these inclusions are equalities if φ is surjective.

Proposition 4.1.4. *The following conditions are equivalent.*

1. *There exists an integer i such that $\mathcal{C}^i(\mathfrak{g}) = \{0\}$.*
2. *There exists an integer j such that $\operatorname{ad} x_1 \circ \operatorname{ad} x_2 \circ \cdots \circ \operatorname{ad} x_j = 0$ for all $x_1, x_2, \ldots x_j \in \mathfrak{g}$.*
3. *There exists a chain $\mathfrak{g} = \mathfrak{g}_1 \supset \mathfrak{g}_2 \supset \cdots \supset \mathfrak{g}_n = \{0\}$ of ideals of $\mathfrak{g}$ such that $[\mathfrak{g}, \mathfrak{g}_i] \subset \mathfrak{g}_{i+1}$ for $i < n$.*

If these conditions are satisfied, we say that $\mathfrak{g}$ is nilpotent.

Proof. The equivalence of 1 and 2 follows from the fact that

$$(\operatorname{ad} x_1 \circ \operatorname{ad} x_2 \circ \cdots \circ \operatorname{ad} x_j)(y) = [x_1, [x_2, \ldots, [x_j, y] \ldots]] \in \mathcal{C}^j(\mathfrak{g})$$

for all $x_1, x_2, \ldots, x_j, y \in \mathfrak{g}$. We have 1. $\Rightarrow$ 3. by taking $\mathfrak{g}_i := \mathcal{C}^i(\mathfrak{g})$. Finally, to show 3. $\Rightarrow$ 1., it is enough to prove $\mathcal{C}^i(\mathfrak{g}) \subset \mathfrak{g}_i$, which is easily proved by induction. □

Proposition 4.1.5. *a) If $\mathfrak{g}$ is nilpotent, then any subalgebra and any quotient of $\mathfrak{g}$ is nilpotent.*

b) Let $\mathfrak{h}$ be a subalgebra of $\mathfrak{g}$ which is contained in the center of $\mathfrak{g}$. Then $\mathfrak{g}$ is nilpotent if and only if $\mathfrak{g}/\mathfrak{h}$ is nilpotent.

Proof. a) is clear. If $\mathfrak{g}/\mathfrak{h}$ is nilpotent, there exists $i \in \mathbb{N}$ such that $\mathcal{C}^i(\mathfrak{g}/\mathfrak{h}) = \{0\}$, so $\mathcal{C}^i(\mathfrak{g}) \subset \mathfrak{h}$. Since $\mathfrak{h}$ is in the center of $\mathfrak{g}$, $\mathcal{C}^{i+1}(\mathfrak{g}) \subset [\mathfrak{g}, \mathfrak{h}] = \{0\}$. □

Proposition 4.1.6. *Let $\mathfrak{g}$ be nilpotent.*

a) If $\mathfrak{g} \neq 0$, then $\mathfrak{z}(\mathfrak{g}) \neq 0$.

b) If $\mathfrak{h}$ is a subalgebra of $\mathfrak{g}$ distinct from $\mathfrak{g}$, then $\mathfrak{n}_{\mathfrak{g}}(\mathfrak{h}) \neq \mathfrak{h}$.

Proof. a) The last nonzero $\mathcal{C}^i(\mathfrak{g})$ is central in $\mathfrak{g}$.

b) Let $j = max\{i : \mathcal{C}^i(\mathfrak{g}) + \mathfrak{h} \neq \mathfrak{h}\}$. Then $[\mathcal{C}^j(\mathfrak{g}) + \mathfrak{h}, \mathfrak{g}] \subset \mathfrak{h}$ which implies $\mathcal{C}^j(\mathfrak{g}) + \mathfrak{h} \subset \mathfrak{n}_{\mathfrak{g}}(\mathfrak{h})$.

□

Proposition 4.1.7. *The following conditions are equivalent.*

1. *There exists an integer i such that $\mathcal{D}^i(\mathfrak{g}) = \{0\}$.*

2. There exists a chain $\mathfrak{g} = \mathfrak{g}_0 \supset \mathfrak{g}_1 \supset \cdots \supset \mathfrak{g}_n = \{0\}$ *of ideals of* $\mathfrak{g}$ *such that* $[\mathfrak{g}_i, \mathfrak{g}_i] \subset \mathfrak{g}_{i+1}$ *for* $0 \leq i \leq n-1$.

If these conditions are satisfied, we say that $\mathfrak{g}$ *is* solvable .

The proof is analogous to the proof of 4.1.4.

Proposition 4.1.8. *a) A nilpotent Lie algebra is solvable.*

b) Subalgebras and quotients of a solvable Lie algebra are solvable.

c) Let $\mathfrak{a}$ *be an ideal of* $\mathfrak{g}$. *Then* $\mathfrak{g}$ *is solvable if and only if* $\mathfrak{a}$ *and* $\mathfrak{g}/\mathfrak{a}$ *are solvable.*

Proof. Parts a) and b) are straightforward. If $\mathfrak{a}$ and $\mathfrak{g}/\mathfrak{a}$ are solvable, then $\mathcal{D}^i(\mathfrak{g}) \subset \mathfrak{a}$ for some integer i and so $\mathcal{D}^{i+j}(\mathfrak{g}) \subset \mathcal{D}^j(\mathfrak{a}) = \{0\}$ for some integer j. So $\mathfrak{g}$ is solvable. □

4.2. The Lie algebra of a linear algebraic group

Let G be a linear algebraic group, $A = C[G]$ its coordinate ring. We consider the set $\operatorname{Der} A$ of derivations of A, i.e C-vector space endomorphisms d of A which moreover satisfy $d(xy) = d(x)y + xd(y)$ for all $x, y \in A$. Clearly $\operatorname{Der} A$ is a C-vector subspace of the C-algebra $End_C A$. It can be checked that the Lie bracket of two derivations is again a derivation; hence $\operatorname{Der} A$ is a Lie algebra. We have seen (Equation (3.1)) that the group G acts on A by translations on the left by $(\lambda_x f)(y) = f(x^{-1}y)$, for $f \in A, x, y \in G$, so we can consider the subspace $\mathfrak{L}(G)$ of left invariant derivations of A, i.e.

$$\mathfrak{L}(G) = \{d \in \operatorname{Der} A : d\lambda_x = \lambda_x d, \forall x \in G\}.$$

The Lie bracket of two derivations in $\mathfrak{L}(G)$ is again in $\mathfrak{L}(G)$; hence $\mathfrak{L}(G)$ is a Lie algebra.

Definition 4.2.1. *The Lie algebra of a linear algebraic group* G is the Lie algebra $\mathfrak{L}(G)$ of left invariant derivations of the coordinate ring $C[G]$ of G.

We shall now compare the Lie algebra $\mathfrak{L}(G)$ with the tangent space of G at the identity element e, $T_eG = T_eG^0$, which has a structure of C-vector space of dimension equal to $\dim G$, as e is a simple point. (See Section 3.1.) To this end we first give an equivalent definition of tangent space of a variety V at a point x in terms of point derivations. We recall that the tangent space of V at x was defined as $(\mathfrak{M}_x/\mathfrak{M}_x^2)^*$.

A *point derivation* at a point x of a variety V is a C-linear map $\delta : \mathcal{O}_x \to C$ such that

$$(4.1) \qquad \delta(fg) = \delta(f)g(x) + f(x)\delta(g).$$

Let $\mathcal{D}_x$ denote the C-vector space of point derivations at x. There is a natural isomorphism $\mathcal{D}_x \simeq (\mathfrak{M}_x/\mathfrak{M}_x^2)^*$. Indeed, if $\delta \in \mathcal{D}_x$, then it vanishes on C and $\mathfrak{M}_x^2$; hence it is determined by its values on $\mathfrak{M}_x/\mathfrak{M}_x^2$. We have then a monomorphism $\mathcal{D}_x \hookrightarrow Tan_x V$. In the other direction, a C-linear map $\mathfrak{M}_x/\mathfrak{M}_x^2 \to C$ defines by composition a map $\mathfrak{M}_x \to C$ and can be extended to $\mathcal{O}_x = C + \mathfrak{M}_x$ by sending the elements in C to 0. Taking into account that, for $f \in \mathcal{O}_x$, $f(x)$ is the image of f under $\mathcal{O}_x \to \mathcal{O}_x/\mathfrak{M}_x \simeq C$, we obtain that the map defined satisfies (4.1).

Now, as point derivations of G at e are already determined by their restriction to $A = C[G]$, we may pass from $\mathfrak{L}(G)$ to $\mathcal{D}_e$ by evaluation at e.

In order to pass from the tangent space of G at e to $\mathfrak{L}(G)$, we shall associate to a vector $\mathbf{x} \in T_eG$ a derivation $*\mathbf{x}$ called *right convolution* by $\mathbf{x}$ and defined by

$$(f * \mathbf{x})(x) = \mathbf{x}(\lambda_{x^{-1}} f), x \in G, f \in C[G],$$

where $\mathbf{x}$ is seen as a point derivation. Let us check that $*\mathbf{x}$ is a left invariant derivation of $A = C[G]$. First, for $f \in A$, let

$$\mu^*(f) = \sum_{i=1}^n f_i \otimes g_i,$$

where $f_i, g_i \in A$ and $\mu^* : A \to A \otimes A$ is the morphism induced by the group law $\mu : G \times G \to G$. For $x, y \in G$, we have $(\lambda_{x^{-1}} f)(y) = f(xy) = \mu^*(f)(x,y) = \sum_{i=1}^n f_i(x) g_i(y)$, so

$$\lambda_{x^{-1}} f = \sum_{i=1}^n f_i(x) g_i.$$

Since $(f * \mathbf{x})(x) = \mathbf{x}(\lambda_{x^{-1}} f)$, it follows that

$$(4.2) \qquad f * \mathbf{x} = \sum_{i=1}^n \mathbf{x}(g_i) f_i.$$

Hence $f * \mathbf{x} \in C[G]$. So right convolution by $\mathbf{x}$ is an endomorphism of $C[G]$.

Now, if $f, g \in A$, we have

$$\begin{aligned}(fg * \mathbf{x})(x) &= \mathbf{x}(\lambda_{x^{-1}}(fg)) \\ &= \mathbf{x}(\lambda_{x^{-1}}(f)\lambda_{x^{-1}}(g)) \\ &= \mathbf{x}(\lambda_{x^{-1}}(f))\lambda_{x^{-1}}(g)(e) + \lambda_{x^{-1}}(f)(e)\mathbf{x}(\lambda_{x^{-1}}(g)) \\ &= \mathbf{x}(\lambda_{x^{-1}}(f))g(x) + f(x)\mathbf{x}(\lambda_{x^{-1}}(g)) \\ &= ((f * \mathbf{x})\, g + f\,(g * \mathbf{x}))(x)\end{aligned}$$

for all $x \in G$, so $*\mathbf{x}$ is a derivation. Now, for $y \in G$,

$$(4.3)\qquad \begin{aligned}(\lambda_y(f * \mathbf{x}))(x) &= (f * \mathbf{x})(y^{-1}x) \\ &= \mathbf{x}(\lambda_{x^{-1}y}f) \\ &= \mathbf{x}(\lambda_{x^{-1}}(\lambda_y f)) \\ &= ((\lambda_y f) * \mathbf{x})(x).\end{aligned}$$

Hence $*\mathbf{x}$ is left invariant.

Proposition 4.2.2. *Let G be a linear algebraic group. The mappings*

$$\theta : \mathfrak{L}(G) \to \mathcal{D}_e$$

defined by $(\theta(d))(f) = (df)(e)$, for $f \in C[G]$, and

$$\eta : \mathcal{D}_e \to \mathfrak{L}(G)$$

defined by $\eta(\mathbf{x}) = *\mathbf{x}$, are mutually inverse isomorphisms of C-vector spaces.

Proof. It is clear that θ and η are C-linear maps. Now, for $d \in \mathfrak{L}(G)$

$$\eta(\theta(d))(f)(x) = \theta(d)(\lambda_{x^{-1}}f) = d(\lambda_{x^{-1}}f)(e) = \lambda_{x^{-1}}(df)(e) = df(x),$$

for all $f \in C[G], x \in G$, so $\eta \circ \theta = Id_{\mathfrak{L}(G)}$ and, for $\mathbf{x} \in \mathcal{D}_e$

$$\theta(\eta(\mathbf{x}))(f) = (f * \mathbf{x})(e) = \mathbf{x}(f)$$

for all $f \in C[G]$, so $\theta \circ \eta = Id_{\mathcal{D}_e}$. □

Using the preceding proposition, we can identify T_eG with $\mathfrak{L}(G)$, so that T_eG inherits a Lie algebra structure. It is in fact more convenient to define the Lie bracket directly on T_eG. Looking at $\mathbf{x}, \mathbf{y} \in T_eG$ as derivations, we can define $\mathbf{x} \otimes \mathbf{y} : A \otimes A \to C$ by $(\mathbf{x} \otimes \mathbf{y})(f \otimes g) = (\mathbf{x}f)(\mathbf{y}g)$ and

$$(4.4)\qquad \mathbf{x} \cdot \mathbf{y} = (\mathbf{x} \otimes \mathbf{y}) \circ \mu^* : A \to C.$$

This product is sent to the product in $\mathfrak{L}(G)$ by η. Indeed, for $f \in A, x \in G$, we have

$$((f * \mathbf{y}) * \mathbf{x})(x) = \mathbf{x}(\lambda_{x^{-1}}(f * \mathbf{y})) = \mathbf{x}((\lambda_{x^{-1}}f) * \mathbf{y}).$$

Let us now set $\mu^*(\lambda_{x^{-1}}f) = \sum f_i \otimes g_i$. Then by (4.2), we have $(\lambda_{x^{-1}}f) * \mathbf{y} = \sum_{i=1}^n \mathbf{y}(g_i) f_i$. Hence

$$((f * \mathbf{y}) * \mathbf{x})(x) = \mathbf{x}(\sum_{i=1}^n \mathbf{y}(g_i) f_i) = \sum_{i=1}^n \mathbf{y}(g_i)\mathbf{x}(f_i).$$

On the other hand

$$(f * (\mathbf{x} \cdot \mathbf{y}))(x) = (\mathbf{x} \cdot \mathbf{y})(\lambda_{x^{-1}}f) = (\mathbf{x} \otimes \mathbf{y})(\mu^*(\lambda_{x^{-1}}f) = \sum_{i=1}^n \mathbf{x}(f_i)\mathbf{y}(g_i).$$

We define then $[\mathbf{x}, \mathbf{y}] = \mathbf{x} \cdot \mathbf{y} - \mathbf{y} \cdot \mathbf{x}$.

Remark 4.2.3. For $\omega_1, \omega_2 \in C[G]^* := \operatorname{Hom}_C(C[G], C)$, we can define similarly $\omega_1 \cdot \omega_2$ by $\omega_1 \cdot \omega_2 = (\omega_1 \otimes \omega_2) \circ \mu^* \in C[G]^*$. Endowed with this product, $C[G]^*$ is an associative C-algebra whose identity element is the evaluation v_e at $e = e_G$. (Recall that the evaluation at an element $x \in G$ is defined by $v_x(f) = f(x)$, for $f \in C[G]$.) Moreover, the map $G \to C[G]^*, x \mapsto v_x$ induces an injective morphism from G into the group of invertible elements of $C[G]^*$.

We saw in the discussion before Proposition 2.2.34 that a morphism of algebraic varieties $\varphi : X \to Y$ induces a C-linear map $d_x\varphi : T_xX \to T_{\varphi(x)}Y$ called the differential mapping of φ at x. If $\varphi : G \to G'$ is a morphism of algebraic groups, we obtain a C-linear map $d_e\varphi : T_eG \to T_eG'$. It is clear that we have $d_e Id_G = Id_{T_eG}$ and $d(\psi \circ \varphi)_e = d\psi_e \circ d\varphi_e$. Considering the definition of the Lie bracket on the tangent spaces given above, the following proposition follows.

Proposition 4.2.4. *If $\varphi : G \to G'$ is a morphism of algebraic groups,*

$$d_e\varphi : T_eG \to T_eG'$$

is a morphism of Lie algebras.

Proof. For $\mathbf{x}, \mathbf{y} \in T_eG$, we have $d_e\varphi(\mathbf{x} \cdot \mathbf{y}) = (\mathbf{x} \otimes \mathbf{y}) \circ \mu^* \circ \varphi^* = (\mathbf{x} \otimes \mathbf{y}) \circ (\varphi^* \otimes \varphi^*) \circ \mu'^* = ((\mathbf{x} \circ \varphi^*) \otimes (\mathbf{y} \circ \varphi^*)) \circ \mu'^* = (\mathbf{x} \circ \varphi^*) \cdot (\mathbf{y} \circ \varphi^*) = d_e\varphi(\mathbf{x}) \cdot d_e\varphi(\mathbf{y})$. □

Proposition 4.2.5. *Let G be an algebraic group, $\mathbf{x}, \mathbf{y} \in \mathfrak{L}(G)$. Then*

$$d\mu(\mathbf{x}, \mathbf{y}) = \mathbf{x} + \mathbf{y}, d\iota(\mathbf{x}) = -\mathbf{x}.$$

Proof. Let $f \in C[G]$ and $\mu^*(f) = \sum_{i=1}^n f_i \otimes g_i$, where $f_i, g_i \in C[G], 1 \leq i \leq n$, and $\mu^* : C[G] \to C[G \times G] \simeq C[G] \otimes C[G]$ is the morphism induced by μ. For $\mathbf{x}, \mathbf{y} \in \mathfrak{L}(G)$, let

$$\theta_{\mathbf{x},\mathbf{y}} : C[G] \otimes C[G] \to C$$

be defined by

$$f \otimes g \mapsto \mathbf{x}(f)g(e) + f(e)\mathbf{y}(g),$$

for $f, g \in C[G]$. Then

$$d\mu(\theta_{\mathbf{x},\mathbf{y}})(f) = \theta_{\mathbf{x},\mathbf{y}}(\mu^*(f)) = \theta_{\mathbf{x},\mathbf{y}}(\sum_{i=1}^n f_i \otimes g_i) = \sum_{i=1}^n (\mathbf{x}(f_i)g_i(e) + f_i(e)\mathbf{y}(g_i)).$$

On the other hand, we have for $x \in G$

$$f(x) = f(ex) = \sum_{i=1}^n f_i(e)g_i(x) = f(xe) = \sum_{i=1}^n f_i(x)g_i(e).$$

Thus

$$f = \sum_{i=1}^n f_i(e)g_i = \sum_{i=1}^n g_i(e)f_i. \tag{4.5}$$

So it is clear that $d\mu(\theta_{\mathbf{x},\mathbf{y}})(f) = \mathbf{x}(f) + \mathbf{y}(f)$.

Now, let $\pi : G \to G \times G$ denote the morphism $x \mapsto (x, \iota(x))$. Then $(\mu \circ \pi)(x) = e$, so $d\mu \circ d\pi = 0$. But $d\pi(\mathbf{x}) = (\mathbf{x}, d\iota(\mathbf{x}))$, so we obtain from the formula for $d\mu$ that $\mathbf{x} + d\iota(\mathbf{x}) = 0$. □

Example 4.2.6. We consider the additive group $\mathbb{G}_a$, whose coordinate ring is the polynomial ring $C[X]$. The Lie algebra $\mathfrak{L}(\mathbb{G}_a)$ is 1-dimensional. Hence as an associative algebra it is commutative, so the Lie bracket is identically zero. Let us see that the derivation $\delta = d/dX$ is left invariant (hence spans $\mathfrak{L}(\mathbb{G}_a)$). It is enough to check it for the polynomial X and left translation by any $x \in \mathbb{G}_a$. We have $\lambda_x \delta X = \lambda_x 1 = 1$ and $\delta \lambda_x X = \delta(X - x) = 1$.

We now consider the multiplicative group $\mathbb{G}_m$, whose coordinate ring is the ring of Laurent polynomials $C[X, X^{-1}]$. The Lie algebra $\mathfrak{L}(\mathbb{G}_m)$ is

1-dimensional. The derivation defined by $\delta X = X$ extends uniquely to the ring $C[X, X^{-1}]$ and is left invariant since $\delta(xX) = x\delta(X)$ for $x \in C^*$.

In the sequel $\mathfrak{g}$ will denote the Lie algebra of the linear algebraic group G.

Example 4.2.7. Let us compute the Lie algebra of $G = \mathrm{GL}(n, C)$. As G is an open subset of the affine space $\mathbb{A}^{n^2}$, its tangent space at e has a canonical basis consisting in the operators $\partial/\partial X_{ij}$ followed by evaluation at the identity matrix. So, a tangent vector $\mathbf{x}$ is determined by the n^2 elements $x_{ij} = \mathbf{x}(X_{ij})$ which can be written as the entries of an $n \times n$ matrix. Then $(\mathbf{x} \cdot \mathbf{y})(X_{ij}) = (\mathbf{x} \otimes \mathbf{y})(\sum_k X_{ik} \otimes X_{kj}) = \sum_k x_{ik}x_{kj}$ gives the usual matrix product. Hence the C-linear map $\mathbf{x} \mapsto (x_{ij})$ from $\mathfrak{g}$ in $M_{n\times n}(C)$ identifies $\mathfrak{g}$ with $\mathfrak{gl}(n, C)$. (The map is injective as only the zero vector annihilates all X_{ij}, and so surjective as both dimensions are equal to n^2.)

If Y is a subvariety of the variety X, the inclusion $i : Y \to X$ induces a monomorphism $d_y i : T_y Y \to T_y X$, for $y \in Y$. Then, if G' is a closed subgroup of the algebraic group G, we can consider the Lie algebra $\mathfrak{g}'$ of G' as a subalgebra of the Lie algebra $\mathfrak{g}$ of G. We shall now determine the Lie algebras of the closed subgroups of $\mathrm{GL}(n, C)$ defined in Example 3.1.4.

Example 4.2.8. **1.** We consider the upper triangular group $T(n, C)$. We can see it as the principal open set of $\mathbb{A}^{n(n-1)/2}$ defined by the nonvanishing of the determinant, so its tangent space in e is the whole ambient affine space. The Lie algebra of $T(n, C)$ is then the Lie algebra $\mathfrak{t}(n, C)$ of all upper triangular matrices.

2. Analogously the diagonal group $D(n, C)$ is the principal open set of $\mathbb{A}^n$ defined by the nonvanishing of the determinant and its Lie algebra is the Lie algebra $\mathfrak{d}(n, C)$ of all diagonal matrices.

3. Let us consider the upper triangular unipotent group $U(n, C)$. It is the closed subset of $\mathbb{A}^{n^2}$ defined by the vanishing of the linear polynomials $X_{ii} - 1$, $1 \leq i \leq n$ and X_{ij}, $1 \leq j < i \leq n$, so the tangent space at e consists on the matrices $(a_{ij})_{1 \leq i,j \leq n}$ with $a_{ij} = 0$ for $j \leq i$. Thus the Lie algebra is $\mathfrak{n}(n, C)$.

4. We now consider the special linear group $SL(n, C)$, which is the zero set of $f(X_{ij}) = \det(X_{ij}) - 1$. We consider the morphism

$$\begin{array}{rccc} \varphi : & \mathrm{GL}(n, C) & \to & \mathrm{GL}(1, C) \\ & A & \mapsto & \det A. \end{array}$$

The tangent space at $e \in \mathrm{GL}(n, C)$ is $\mathbb{A}^{n^2}$. If we look at the tangent space as the C-vector space of point derivations, an element $a =$

$(a_{11}, \dots, a_{nn}) \in \mathbb{A}^{n^2}$ corresponds to $\sum_{i=1}^n \sum_{j=1}^n a_{ij}\partial/\partial X_{ij}$ followed by evaluation at e. Then

$$(4.6) \quad d\varphi_e(a) = \sum_{k=1}^n \sum_{l=1}^n a_{kl} \frac{\partial \det(X_{ij})}{\partial X_{kl}}(Id) = a_{11} + a_{22} + \dots + a_{nn} \in \mathbb{A}^1.$$

Indeed, as $\partial \det(X_{ij})/\partial X_{kl}$ is the minor of the matrix (X_{ij}) obtained by deleting the kth row and the lth column, by evaluating at the identity matrix we obtain 1 for $k = l$ and 0 otherwise. So the Lie algebra of $\mathrm{SL}(n, C)$ is the Lie algebra $\mathfrak{sl}(n, C)$ of $n \times n$ matrices with trace equal to zero. Equation 4.6 also shows that the differential of the morphism $\mathrm{GL}(n, C) \to \mathbb{G}_m$ associating to each matrix its determinant is the map $\mathfrak{gl}(n, C) \to C$ associating to each matrix its trace.

Let H be a closed subgroup of a linear algebraic group G. The inclusion $i : H \to G$ induces an epimorphism $i^* : C[G] \to C[G]/I \simeq C[H]$, for I the ideal of $C[G]$ vanishing on H. Therefore di_e identifies T_eH with the subspace of T_eG consisting of those $\mathbf{x}$ for which $\mathbf{x}(I) = 0$. We saw in Lemma 3.4.1 a characterization of closed subgroups by means of right translation. We shall now see a characterization of the Lie algebra of a closed subgroup by means of right convolution.

Lemma 4.2.9. *Let G be a linear algebraic group, H a closed subgroup of G, $I = \mathcal{I}(H)$. Let $\mathfrak{h}$ be the Lie algebra of H. We have*

$$\mathfrak{h} = \{\mathbf{x} \in \mathfrak{g} : I * \mathbf{x} \subset I\} = \{\mathbf{x} \in \mathfrak{g} : \mathbf{x}(I) = 0\}.$$

Proof. For $\mathbf{x} \in \mathfrak{h}, f \in I, x \in H$, we have $(f * \mathbf{x})(x) = \mathbf{x}(\lambda_{x^{-1}} f) = 0$, since $\lambda_{x^{-1}} f$ belongs to I. So $f * \mathbf{x} \in I$.

In the other direction, let $\mathbf{x} \in \mathfrak{g}$ satisfy $I * \mathbf{x} \subset I$. If $f \in I$, then $(f * \mathbf{x})(e) = \mathbf{x}(\lambda_{e^{-1}} f) = \mathbf{x}(f)$. By hypothesis, $f * \mathbf{x} \in I$; hence $(f * \mathbf{x})(e) = 0$, so $\mathbf{x} \in \mathfrak{h}$.

Finally if $f \in I$ and $\mathbf{x} \in \mathfrak{g}$ satisfy $f * \mathbf{x} \in I$, then $\mathbf{x}(f) = (f * \mathbf{x})(e) = 0$. Conversely if $\mathbf{x}(f) = 0$ for all $f \in I$, then since $\lambda_{x^{-1}} f \in I$ for $x \in H$, we have $(f * \mathbf{x})(x) = \mathbf{x}(\lambda_{x^{-1}} f) = 0$, and so $f * \mathbf{x} \in I$. □

We shall now consider the differential of right translation. We saw in Section 3.4 that a linear algebraic group G acts on $C[G]$ by left and right translations. For $f \in C[G], x, y \in G$, we have $(\lambda_x f)(y) = f(x^{-1}y), (\rho_x f)(y) = f(yx)$. We then obtain group morphisms $\lambda, \rho : G \to \mathrm{GL}(C[G])$ which cannot be considered as morphisms of algebraic groups, since $C[G]$ is infinite dimensional. However, by Lemma 3.4.2, $C[G]$ is the union of finite dimensional subspaces stable under all ρ_x (or all λ_x). Let E be a subspace of $C[G]$ stable under all ρ_x and let $\varphi : G \times E \to E$ be the corresponding action. If $(f_1, \ldots, f_n)$ is a basis of E, then we saw in 3.4.2 and the proof of Theorem 3.4.3 that $\rho_x f_i = \sum_j m_{ij}(x) f_j$, for $\varphi^*(f_i) = \sum_j m_{ij} \otimes f_j$, i.e. that $(m_{ij})_{1 \le i,j \le n}$ is the matrix of $\rho_x|E$ in the basis $(f_1, \ldots, f_n)$. We then obtain a morphism $\psi : G \to \mathrm{GL}(n, C), x \mapsto (m_{ij})$. Notice that the subspace of $C[G]$ spanned by the m_{ij} includes E and all its left translates, as we have

$$(4.7) \qquad (\lambda_{x^{-1}} f_i)(y) = f_i(xy) = \sum_j m_{ij}(y) f_j(x).$$

If we denote by X_{ij} the coordinate functions on $\mathrm{GL}(n, C)$, we have $\psi^*(X_{ij})(x) = m_{ij}$; hence, for $\mathbf{x} \in \mathfrak{g}$, $d_e\psi(\mathbf{x}) = (\mathbf{x}(m_{ij}))$. On the other hand, consider the action of $\mathbf{x}$ on the given basis of E by right convolution: $(f_i * \mathbf{x})(x) = \mathbf{x}(\lambda_{x^{-1}} f_i) = \mathbf{x}(\sum_j f_j(x) m_{ij}) = \sum_j f_j(x)\mathbf{x}(m_{ij})$, by (4.7). So, $\mathbf{x}$ leaves E stable and has matrix $\mathbf{x}(m_{ij})$ relative to the basis $(f_1, \ldots, f_n)$. We have then obtained the following result.

Proposition 4.2.10. *Acting by right convolution, $\mathfrak{g}$ leaves stable every subspace of $C[G]$ stable under right translation by G. On a finite dimensional G-stable subspace, the differential of right translation is right convolution.*

We could replace right by left, with left convolution defined by

$$(\mathbf{x} * f)(x) = \mathbf{x}(\rho_x f)$$

for $\mathbf{x} \in \mathfrak{g}, f \in C[G], x \in G$.

4.3. Decomposition of algebraic groups

Let $x \in \operatorname{End} V$, for V a finite dimensional vector space over C. Then x is *nilpotent* if $x^n = 0$ for some n (equivalently if 0 is the only eigenvalue of x). At the other extreme, x is called *semisimple* if the minimal polynomial of x has distinct roots (equivalently if x is diagonalizable over C). From the fact that a square matrix with entries in C is conjugated to one in Jordan canonical form, we obtain the Jordan additive decomposition stated in the next lemma. (See [**L**].)

Lemma 4.3.1. *Let $x \in \operatorname{End} V$.*

a) *There exist unique $x_s, x_n \in \operatorname{End} V$ such that x_s is semisimple, x_n is nilpotent and $x = x_s + x_n$.*

b) *There exist polynomials $P(T), Q(T) \in C[T]$, without constant term such that $x_s = P(x), x_n = Q(x)$. Hence x_s and x_n commute with any endomorphism of V which commutes with x; in particular, they commute with each other.*

c) *If $W_1 \subset W_2$ are subspaces of V, and x maps W_2 into W_1, then so do x_s and x_n.*

d) *Let $y \in \operatorname{End} V$. If $xy = yx$, then $(x+y)_s = x_s+y_s$ and $(x+y)_n = x_n+y_n$.*
□

If $x \in \mathrm{GL}(V)$, its eigenvalues are nonzero, and so x_s is also invertible. We can write $x_u := 1+x_s^{-1}x_n$ and then we obtain $x = x_s+x_n = x_s(1+x_s^{-1}x_n) = x_s \cdot x_u$. We call an element in $\mathrm{GL}(V)$ *unipotent* if it is the sum of the identity and a nilpotent endomorphism or, equivalently, if 1 is its unique eigenvalue. For $x \in \mathrm{GL}(V)$, the Jordan multiplicative decomposition $x = x_s \cdot x_u$, with x_s semisimple, x_u unipotent, is unique. Clearly the only element in $\mathrm{GL}(V)$ which is both semisimple and unipotent is identity. From Lemma 4.3.1, we further obtain

Lemma 4.3.2. *Let $x \in \mathrm{GL}(V)$.*

a) *There exist unique $x_s, x_u \in \mathrm{GL}(V)$ such that x_s is semisimple, x_u is unipotent, $x = x_s x_u$ and $x_s x_u = x_u x_s$.*

b) *x_s and x_u commute with any endomorphism of V which commutes with x.*

c) *If W is a subspace of V stable under x, then W is stable under x_s and x_u.*

d) *Let $y \in \mathrm{GL}(V)$. If $xy = yx$, then $(xy)_s = x_s y_s$ and $(xy)_u = x_u y_u$.* □

It is sometimes useful to allow V to be infinite dimensional, even though the notions "semisimple" and "unipotent" do not carry over directly to this case. If $x \in \mathrm{GL}(V)$ and if V is the union of finite dimensional subspaces stable under x, then the decompositions $x|W = (x|W)_s(x|W)_u$ exist for all such subspaces W. Moreover, the restriction of a semisimple (resp. unipotent) endomorphism to an intersection $W \cap W'$ is of the same type, so we can patch together the $(x|W)_s$ (resp. $(x|W)_u$) to obtain invertible endomorphisms of V, of which x is the product. These may again be denoted x_s, x_u and called the Jordan parts of x. It is important to observe, using Lemma 4.3.2 c), that x_s and x_u leave stable every subspace of V, finite dimensional or not, which is stable under x. Similar remarks apply to the additive Jordan decomposition.

If G is an arbitrary subgroup of $\mathrm{GL}(n, C)$, G does not necessarily contain the semisimple and unipotent part of each of its elements. However, it is so for closed subgroups. Applying the membership criterion 3.4.1, given $x \in G$, we have to see that ρ_{x_s} and ρ_{x_u} leave stable the ideal $\mathcal{I}(G) \subset C[\mathrm{GL}(n, C]$.

We shall see that ρ_{x_s} and ρ_{x_u} are the semisimple and unipotent parts of ρ_x. The preceding question will then follow by Lemma 4.3.2 c).

Proposition 4.3.3. *Let* $G = \mathrm{GL}(n,C), \mathfrak{g} = \mathfrak{gl}(n,C)$. *If* $x \in G$ *(resp.* $\mathbf{x} \in \mathfrak{g}$*), then* ρ_x *(resp* $*\mathbf{x}$*) has Jordan decomposition* $\rho_{x_s}\rho_{x_u}$ *(resp* $*\mathbf{x}_s + *\mathbf{x}_n$*).*

Proof. Since $C[G]$ is the union of finite dimensional subspaces stable under all ρ_x (hence under all $*\mathbf{x}$ by Proposition 4.2.10), Jordan decompositions do exist. Moreover $\rho_x = \rho_{x_s x_u} = \rho_{x_s}\rho_{x_u}$, $*\mathbf{x} = *(\mathbf{x}_s + \mathbf{x}_n) = *\mathbf{x}_s + *\mathbf{x}_n$ and the operators commute. So it will be enough to show that ρ_{x_s} and $*\mathbf{x}_s$ are semisimple, that ρ_{x_u} is unipotent and $*\mathbf{x}_n$ is nilpotent.

The coordinate ring $C[G]$ is the ring of polynomials in n^2 indeterminates X_{ij} localized in the multiplicative system of the powers of $d = \det(X_{ij})$. Let us see that $C[X_{ij}]$ is stable under both right translation and right convolution. For $x, y \in G$, we have $(\rho_x X_{ij})(y) = X_{ij}(yx) = \sum_h y_{ih}x_{hj} = \sum_h X_{ih}(y)x_{hj}$; hence $\rho_x X_{ij} = \sum_h x_{hj}X_{ih} \in C[X_{ij}]$. Similarly, we get $\lambda_{x^{-1}}X_{ij} = \sum_h x_{ih}X_{hj}$. From this we obtain for $\mathbf{x} \in \mathfrak{g}, y \in G$,

$$(X_{ij}*\mathbf{x})(y) = \mathbf{x}(\lambda_{y^{-1}}X_{ij}) = \mathbf{x}(\sum_h y_{ih}X_{hj}) = \sum_h y_{ih}\mathbf{x}(X_{hj}) = \sum_h \mathbf{x}_{hj}X_{ih}(y).$$

We shall now describe how G and $\mathfrak{g}$ act on d. For $x, y \in G$, $(\rho_x d)(y) = \det(yx) = \det y \cdot \det x$, so $\rho_x d = \det x \cdot d$. This shows that the vector space spanned by d is G-stable, hence $\mathfrak{g}$-stable by Proposition 4.2.10 and $d * \mathbf{x} = \mathrm{Tr}(\mathbf{x})d$, as the differential of the determinant is the trace. (See Example 4.2.8 **4.**) From this we see how to describe the action of ρ_x or $*\mathbf{x}$ on $C[G]$ once the action on $C[X_{ij}]$ is known.

In particular, as d is an eigenvector of ρ_x in any case, we have that if $\rho_x|C[X_{ij}]$ is semisimple, then ρ_x is semisimple. If $\rho_x|C[X_{ij}]$ is unipotent, then its eigenvalue $\det x$ for d must be 1, so ρ_x is unipotent. Similarly, if $*\mathbf{x}|C[X_{ij}]$ is semisimple (resp. nilpotent), then so is $*\mathbf{x}$. Therefore it will be enough to consider the actions of $\rho_x, *\mathbf{x}$ on $C[X_{ij}]$ in place of $C[G]$.

Let $E = End\,V$, where $V = C^n$, and regard G as the subset $\mathrm{GL}(V)$ of E, while $\mathfrak{g} = \mathfrak{gl}(V) = E$. The algebra $C[X_{ij}]$ may be identified with the symmetric algebra $S(E^*)$ on the dual space E^* of E. If $x \in E$, define an endomorphism $r_x : E \to E$ by $r_x(y) = yx$, and let $r_x^* : E^* \to E^*$ be the dual endomorphism. Then we have $r_x^*(X_{ij}) = \sum_{h=1}^n x_{hj}X_{ih}$ and comparing with the formulas obtained above for ρ_x and $*\mathbf{x}$, we see that ρ_x (resp. $*\mathbf{x}$) is just the canonical extension of r_x^* (resp. $r_{\mathbf{x}}^*$) to an automorphism (resp. derivation) of $S(E^*)$. So it just has to be verified that the property of being semisimple or unipotent (resp. semisimple or nilpotent) is preserved at each step when we pass from x to r_x to r_x^* to ρ_x (resp. from $\mathbf{x}$ to $r_{\mathbf{x}}$ to $r_{\mathbf{x}}^*$ to $*\mathbf{x}$).

The passage from r_x to r_x^* poses no problem, neither does going from the action on E^* to the action on $S(E^*)$. It remains to treat the passage from x to r_x, for $x \in E$.

If $x \in E$ is semisimple, choose a basis $(v_1, \dots, v_n)$ of V consisting of eigenvectors, so that $x(v_i) = a_i v_i$, for some $a_i \in C$. Take in E the basis e_{ij} defined by $e_{ij}(v_k) = \delta_{jk} v_i$, then $r_x(e_{ij}) = a_j e_{ij}$, so r_x is semisimple. If $x \in E$ is nilpotent, say $x^t = 0$, then $r_x^t(y) = yx^t = 0$ for all $y \in E$, so r_x is nilpotent. If x is unipotent, then $n = x - 1$ is nilpotent and $r_x = 1 + r_n$ with r_n nilpotent, so r_x is unipotent. □

We shall now see that we can consider Jordan decomposition for elements in any affine algebraic group.

Proposition 4.3.4. *Let G be an affine algebraic group.*

a) If $x \in G$, there exist unique elements $s, u \in G$ such that $x = su$, s and u commute, ρ_s is semisimple, ρ_u is unipotent. Then we call s and u the semisimple part *of x and the* unipotent part *of x, respectively and denote them x_s and x_u.*

b) If $\mathbf{x} \in \mathfrak{g}$, there exist unique elements $\mathfrak{s}, \mathfrak{n} \in \mathfrak{g}$ such that $\mathbf{x} = \mathfrak{s} + \mathfrak{n}$, $[\mathfrak{s}, \mathfrak{n}] = 0$, $*\mathfrak{s}$ is semisimple, $*\mathfrak{n}$ is nilpotent. Then we call $*\mathfrak{s}$ and $*\mathfrak{n}$ the semisimple part *and the* nilpotent part *of $\mathbf{x}$, respectively and denote them $\mathbf{x}_s$ and $\mathbf{x}_n$.*

c) If $\varphi : G \to G'$ is a morphism of algebraic groups, then $\varphi(x)_s = \varphi(x_s)$, $\varphi(x)_u = \varphi(x_u)$, $d\varphi(\mathbf{x})_s = d\varphi(\mathbf{x}_s)$, $d\varphi(\mathbf{x})_n = d\varphi(\mathbf{x}_n)$ for all $x \in G$, $\mathbf{x} \in \mathfrak{g}$.

Proof. We may embed G as a closed subgroup of some $\mathrm{GL}(n, C)$. (See Theorem 3.4.3.) If I is the ideal in $C[\mathrm{GL}(n, C)]$ defining G, then the criterion for $x \in \mathrm{GL}(n, C)$ (resp. $\mathbf{x} \in \mathfrak{gl}(n, C)$) to be in G (resp. $\mathfrak{g}$) is that ρ_x (resp $*\mathbf{x}$) stabilize I. (See Lemmas 3.4.1, 4.2.9.) Now let $x \in G$ and $x = su$ its Jordan decomposition in $\mathrm{GL}(n, C)$. By Proposition 4.3.3, $(\rho_x)_s = \rho_s$ and $(\rho_x)_u = \rho_u$. By the remarks preceding Proposition 4.3.3, both ρ_s and ρ_u stabilize I, so $s, u \in G$. For $\mathbf{x} \in \mathfrak{g}$, by an analogous argument, we obtain b).

Next consider $\varphi : G \to G'$ which factors into two morphisms: the epimorphism $G \twoheadrightarrow \varphi(G)$ followed by the inclusion $\varphi(G) \hookrightarrow G'$. It suffices to treat each of these cases separately. In case φ is an epimorphism, right translation by $\varphi(x)$ is essentially the restriction of ρ_x to $C[G']$, viewed as a subring of $C[G]$ and similarly for right convolution. But the restriction of a semisimple (resp. unipotent) operator to a subspace is again of the same type. It follows that $\rho_{\varphi(x)} = \rho_{\varphi(x_s)} \rho_{\varphi(x_u)}$ is the Jordan decomposition of $\rho_{\varphi(x)}$, hence that $\varphi(x_s) = \varphi(x)_s, \varphi(x_u) = \varphi(x)_u$ and similarly for $d\varphi(\mathbf{x})$. In case φ is an inclusion, the situation is just like the one considered above (G viewed as a closed subgroup of $\mathrm{GL}(n, C)$). □

The proposition shows that in any affine algebraic group G, the subsets

$$G_s = \{x \in G : x = x_s\} \quad \text{and} \quad G_u = \{x \in G : x = x_u\}$$

are intrinsically defined and intersect in e. Similarly

$$\mathfrak{g}_s = \{\mathbf{x} \in \mathfrak{g} : \mathbf{x} = \mathbf{x}_s\} \quad \text{and} \quad \mathfrak{g}_n = \{\mathbf{x} \in \mathfrak{g} : \mathbf{x} = \mathbf{x}_n\}$$

intersect in 0. Part c) of the proposition ensures that morphisms of algebraic groups and their differentials preserve these sets. Moreover G_u and $\mathfrak{g}_n$ are closed sets ($\mathfrak{g}$ being given the topology of an affine space). To see this, just observe that the set of all unipotent (resp. nilpotent) matrices in $\mathrm{GL}(n, C)$ (resp. in $\mathfrak{gl}(n, C)$) is closed, being the zero set of the polynomials implied by $(x - 1)^n = 0$ (resp. $\mathbf{x}^n = 0$). By contrast, G_s is not in general a closed subset of G. However, see Theorem 4.3.6.

Let us denote by $\mathcal{T}(n, C)$ (resp. $\mathcal{D}(n, C)$) the ring of all upper triangular (resp. all diagonal) matrices in $M(n, C)$. A subset M of $M(n, C)$ is said to be *triangularizable* (resp. *diagonalizable*) if there exists $x \in \mathrm{GL}(n, C)$ such that $xMx^{-1} \subset \mathcal{T}(n, C)$ (resp. $\mathcal{D}(n, C)$).

Lemma 4.3.5. *If $M \subset M(n, C)$ is a commuting set of matrices, then M is triangularizable. If M has a subset N consisting of diagonalizable matrices, N can be diagonalized at the same time.*

Proof. Let $V = C^n$ and proceed by induction on n. If $x \in M, \lambda \in C$, the subspace $W = \mathrm{Ker}(x - \lambda I)$ is evidently stable under the endomorphisms of V which commute with x; hence it is stable under M. Unless M consists of scalar matrices (then we are done), it is possible to choose x and λ such that $0 \neq W \neq V$. By induction, there exists a nonzero $v_1 \in W$ such that Cv_1 is M-stable. Applying the induction hypothesis next to the induced action of M on V/Cv_1, we obtain $v_2, \ldots, v_n \in V$ completing the basis for V, such that M stabilizes each subspace $Cv_1 + \cdots + Cv_i$ $(1 \leq i \leq n)$. The transition from the canonical basis of V to $(v_1, \ldots v_n)$ therefore triangularizes M.

Now if N does not already consist of scalar matrices, we can choose x above to lie in N. Since x is diagonalizable, $V = W \oplus W'$, where the sum W' of remaining eigenspaces of x is nonzero. As before, both W and W' are M-stable. The induction hypothesis allows us to choose bases of W and W' which triangularize M while simultaneously diagonalizing N. □

Theorem 4.3.6. *Let G be a commutative linear algebraic group. Then G_s, G_u are closed subgroups, connected if G is connected, and the product map $\varphi : G_s \times G_u \to G$ is an isomorphism of algebraic groups. Moreover $\mathfrak{L}(G_s) = \mathfrak{g}_s$ and $\mathfrak{L}(G_u) = \mathfrak{g}_n$.*

Proof. As G is commutative, by Lemma 4.3.2 d), G_s and G_u are subgroups of G. We already observed in the remarks following Proposition 4.3.4 that G_u is closed. As G is commutative, by Lemma 4.3.2 a), φ is a group isomorphism. Now embed G in some $\mathrm{GL}(n, C)$. By Lemma 4.3.5, we may assume that $G \subset \mathrm{T}(n, C)$ and $G_s \subset \mathrm{D}(n, C)$. This forces $G_s = G \cap \mathrm{D}(n, C)$, so G_s is also closed. Moreover, φ is a morphism of algebraic groups.

It has to be shown that the inverse map is a morphism of algebraic groups. To this end, it suffices to show that $x \mapsto x_s$ and $x \mapsto x_u$ are morphisms. Since, $x_u = x_s^{-1}x$, if the first map is a morphism, the second will also be. Now, if $x \in G$, x_s is the diagonal part of x; hence $x \mapsto x_s$ is a morphism. Furthermore, if G is connected, so are G_s and G_u since there are homomorphic images of G.

The chosen embedding of G in $\mathrm{T}(n, C)$ shows also that $\mathfrak{L}(G_s) \subset \mathfrak{d}(n, C)$, $\mathfrak{L}(G_u) \subset \mathfrak{n}(n, C)$. Therefore $\mathfrak{L}(G_s) \subset \mathfrak{g}_s$ and $\mathfrak{L}(G_u) \subset \mathfrak{g}_n$. But φ is an isomorphism, so $\mathfrak{L}(G_s) \oplus \mathfrak{L}(G_u) = \mathfrak{g}$. Since also $\mathfrak{g} = \mathfrak{g}_s + \mathfrak{g}_n$ (with uniqueness of expression), both inclusions are equalities. □

4.4. Solvable algebraic groups

For a group G, we denote by $[x, y]$ the commutator $xyx^{-1}y^{-1}$ of two elements $x, y \in G$. If A and B are two subgroups of G we denote by $[A, B]$ the subgroup generated by all commutators $[a, b]$ with $a \in A, b \in B$. In particular $[G, G]$ is called the *derived subgroup of* G. The identity

$$z[x, y]z^{-1} = [zxz^{-1}, zyz^{-1}] \tag{4.8}$$

shows that $[A, B]$ is normal in G if both A and B are normal in G.

We denote by $Z(G)$ the *center* of a group G, i.e.

$$Z(G) = \{x \in G : xy = yx \,, \forall y \in G\}.$$

Lemma 4.4.1. *a) If the index $[G : Z(G)]$ is finite, then $[G, G]$ is finite.*

b) Let A, B be normal subgroups of G, and suppose the set

$$S = \{[x, y] : x \in A, y \in B\}$$

is finite. Then $[A, B]$ is finite.

Proof. a) Let $n = [G : Z(G)]$ and let S be the set of all commutators in G. Then S generates $[G, G]$. For $x, y \in G$, it is clear that $[x, y]$ depends only on the cosets of x, y modulo $Z(G)$; in particular, $Card\, S \leq n^2$. Given a product of commutators, any two of them can be made adjacent by suitable conjugation, e.g. $[x_1, y_1][x_2, y_2][x_3, y_3] = [x_1, y_1][x_3, y_3][z^{-1}x_2z, z^{-1}y_2z]$, where $z = [x_3, y_3]$. Therefore, it is enough to show that the $(n+1)$th power of an element of S is the product of n elements of S, in order to conclude that each element of $[G, G]$ is the product of at most n^3 factors from S. This in turn will force $[G, G]$ to be finite. Now $[x, y]^n \in Z(G)$ and so we can write $[x, y]^{n+1} = y^{-1}[x, y]^n y[x, y] = y^{-1}[x, y]^{n-1}[x, y^2]y$, and the last expression can be written as a product of n commutators by using identity (4.8).

b) We can assume that $G = AB$. Taking into account identity (4.8), we see that G acts on S by inner automorphisms. If H is the kernel of the resulting morphism from G in the group $Sym(S)$ of permutations of S, then clearly, H is a normal subgroup of finite index in G. Moreover, H centralizes $E = [A, B]$. It follows that $H \cap E$ is central in E and of finite index. By a), $[E, E]$ is finite (as well as normal in G, since $E \lhd G$). So we can replace G by $G/[E, E]$, i.e. we can assume that E is abelian.

Now the commutators $[x, y], x \in A, y \in E$, lie in E and commute with each other. As E is abelian and normal in G, $[x, y]^2 = (xyx^{-1})^2y^{-2} = [x, y^2]$ is another such commutator. This clearly forces $[A, E]$ to be finite (as well as normal in G). Replacing G by $G/[A, E]$, we may further assume that A centralizes E. This implies that the square of an arbitrary commutator is again a commutator. It follows that $[A, B]$ is finite. □

Proposition 4.4.2. *Let A, B be closed subgroups of an algebraic group G.*

a) If A is connected, then $[A, B]$ is closed and connected. In particular, $[G, G]$ is connected if G is.

b) If A and B are normal in G, then $[A, B]$ is closed (and normal) in G. In particular, $[G, G]$ is always closed.

Proof. a) For each $b \in B$, we can define the morphism $\varphi_b : A \to G$ by $a \mapsto [a, b]$. Since A is connected and $\varphi_b(e) = e$, by Proposition 3.2.3, the group generated by all $\varphi_b(A), b \in B$ is closed and connected and this is exactly $[A, B]$.

b) It follows from part a) that $[A^0, B]$ and $[A, B^0]$ are closed, connected (as well as normal) subgroups of G, so by Proposition 3.3.3 their product E is a closed normal subgroup of G. To show that $[A, B]$ is closed, it therefore suffices to show that E has finite index in $[A, B]$, which is a purely group-theoretic question. In the abstract group G/E, the image of A^0 (resp. B^0)

centralizes the image of B (resp. A). Since the indices $[A : A^0]$ and $[B : B^0]$ are finite, this implies that there are only finitely many commutators in G/E constructible from the images of A and B. Lemma 4.4.1 b) then guarantees that $[A, B]/E$ is finite. □

For an abstract group G, we define the *derived series* D^iG inductively by

$$D^0G = G\,,\ D^{i+1}G = [D^iG, D^iG], i \geq 0.$$

We say that G is *solvable* if its derived series terminates in e.

If G is an algebraic group, $D^1G = [G, G]$ is a closed normal subgroup of G, connected if G is, by Proposition 4.4.2. By induction the same holds true for all D^iG. If G is a connected solvable algebraic group of positive dimension, we have $\dim[G, G] < \dim G$.

It is easy to see that an algebraic group G is solvable if and only if there exists a chain of closed subgroups $G = G_0 \supset G_1 \supset \cdots \supset G_n = e$ such that $G_i \lhd G_{i-1}$ and G_{i-1}/G_i is abelian, for $i = 1, \ldots, n$.

The following results from group theory are well known. (See e.g. [**Sc**].)

Proposition 4.4.3. *a) Subgroups and homomorphic images of a solvable group are solvable.*

b) If N is a normal solvable subgroup of G for which G/N is solvable, then G itself is solvable.

c) If A and B are normal solvable subgroups of G, so is AB. □

The following lemma will be used in the characterization of Liouville extensions.

Lemma 4.4.4. *Let G be an algebraic group, H a closed subgroup of G. Suppose that H is normal in G and G/H is abelian. Suppose further that the identity component H^0 of H is solvable. Then the identity component G^0 of G is solvable.*

Proof. We have $[G, G] \subset H$; whence $[G^0, G^0] \subset H$. By Proposition 4.4.2, $[G^0, G^0]$ is connected. Hence $[G^0, G^0] \subset H^0$. By hypothesis H^0 is solvable, whence $[G^0, G^0]$ is solvable, and then G^0 is solvable. □

Example 4.4.5. We consider the groups $\mathrm{T}(n, C)$ and $\mathrm{U}(n, C)$. We know by Corollary 3.2.5 that they are connected. We shall now see that they are solvable. Write $T = \mathrm{T}(n, C)$, $U = \mathrm{U}(n, C)$, $D = D(n, C)$. First, since the diagonal entries in the product of two upper triangular matrices are just the respective products of diagonal entries it is clear that $[T,T] \subset U$. Now we know that U is generated by the subgroups U_{ij} with $i<j$, each of them isomorphic to $\mathbb{G}_a$. (See the proof of Corollary 3.2.5.) By Proposition 4.4.2, we have that $[D,U_{ij}] \subset U_{ij}$ is closed and connected, and clearly this group is nontrivial. Then $U_{ij} \subset [D,U_{ij}] \subset [T,T]$. We have then proved $[T,T]=U$.

Now we want to prove that U is solvable. This will imply that T is solvable as well. Let us denote by $\mathcal{T}$ the full set of upper triangular matrices viewed as a ring. The subset $\mathcal{N}$ of matrices with 0 diagonal is a 2-sided ideal of $\mathcal{T}$. So each ideal power $\mathcal{N}^h$ is again a two-sided ideal. For an element $u \in U$, such that $u = 1+a$, with $a \in \mathcal{N}$, we have $(1+a)^{-1} = 1-a+a^2-a^3+\cdots+(-1)^{n-1}a^{n-1}$. If we set $U_h = 1+\mathcal{N}^h$, we obtain $[U_h,U_l] \subset U_{h+l}$. In particular, U is solvable.

The next theorem establishes that the connected solvable subgroups of $\mathrm{GL}(n,C)$ are exactly the conjugate subgroups of $\mathrm{T}(n,C)$.

Theorem 4.4.6 (Lie-Kolchin). *Let G be a connected solvable subgroup of* $\mathrm{GL}(n,C)$, $n \geq 1$. *Then G is triangularizable.*

Proof. Let $V = C^n$. Let us first assume that G is reducible, i.e. that V admits a nontrivial invariant subspace W. If a basis of W is extended to a basis of V, the matrices representing G have the form

$$\begin{pmatrix} \varphi(x) & * \\ 0 & \psi(x) \end{pmatrix}.$$

The morphism $x \mapsto \varphi(x)$ is a morphism of algebraic groups. As G is connected, $\varphi(G) \subset \mathrm{GL}(W)$ is also connected as well as solvable (Proposition 4.4.3 a)). By induction on n, $\varphi(G)$ can be triangularized. Analogously, we obtain that $\psi(G)$ can be triangularized as well. We then obtain the triangularization for G itself. We may then assume that G is irreducible.

By Proposition 4.4.2, the commutator subgroup $[G,G]$ of G is connected, so by induction on the length of the derived series, we can assume that $[G,G]$ is in triangular form.

Let V_1 be the subspace of V generated by all common eigenvectors of $[G,G]$. We have $V_1 \neq 0$, since the triangular form of $[G,G]$ yields at least one common eigenvector. Now, for each $x \in G$, $y \in [G,G]$, we have $x^{-1}yx \in [G,G]$; hence for each $v \in V_1$, $(x^{-1}yx)(v) = \lambda v$, for some $\lambda \in C$, equivalently $y(xv) = \lambda xv$. So, V_1 is G-stable. Since G is irreducible, $V_1 = V$, which means that $[G,G]$ is in diagonal form.

Now, any element in $[G,G]$ is a diagonal matrix. Its conjugates in G are again in $[G,G]$, hence also diagonal. The only possible conjugates are then obtained by permuting the eigenvalues. Hence each element in $[G,G]$ has a finite conjugacy class. By Proposition 3.2.1c), $[G,G]$ lies in the center of G.

Assume that there is a matrix $y \in [G,G]$ which is not a scalar. Let λ be an eigenvalue of y, and W the corresponding eigenspace. Since y commutes with all elements in G, W is G-invariant; hence $W = V$, $y = \lambda \cdot 1$.

Since $[G,G]$ is the commutator subgroup of G, its elements have determinant 1. Hence the diagonal entries must be n-th roots of unity. There are only a finite number of these, so $[G,G]$ is finite. But by Proposition 4.4.2, $[G,G]$ is connected, then $[G,G] = 1$, which means that G is commutative. The result then follows from Lemma 4.3.5. □

For an abstract group G, we define the *central descending series* C^iG inductively by

$$C^1G = G\,,\ C^{i+1}G = [G, C^iG], i \geq 1.$$

We say that G is *nilpotent* if its central descending series terminates in e.

4.5. Correspondence between algebraic groups and Lie algebras

In this section we study the correspondence between closed subgroups of an affine algebraic group and Lie subalgebras of its Lie algebra. From now on we are assuming all algebraic groups to be affine. We are particularly interested in the relation between solvability and nilpotency of an algebraic group and its Lie algebra. For an algebraic group G, we denote by $\mathfrak{g}$ its Lie algebra.

For $x \in G$, denote by i_x the inner automorphism of G defined by $y \mapsto xyx^{-1}$. Let $\mathrm{Ad}_G(x)$ or $\mathrm{Ad}(x)$ be the differential of i_x. It is an automorphism of the Lie algebra $\mathfrak{g}$. Since $i_{xy} = i_x \circ i_y$, for $x, y \in G$, we deduce that $\mathrm{Ad}(xy) = \mathrm{Ad}(x) \circ \mathrm{Ad}(y)$. Hence the map

$$\mathrm{Ad} : G \to \mathrm{GL}(\mathfrak{g})\,,\ x \mapsto \mathrm{Ad}(x)$$

defines a representation of G, called the *adjoint representation* of G.

Lemma 4.5.1. *For $x \in G, \mathbf{x} \in \mathfrak{L}(G)$, we have*

$$\mathrm{Ad}(x)(\mathbf{x}) = v_x \cdot \mathbf{x} \cdot v_{x^{-1}}.$$

Proof. Let $f \in C[G], \mu^*(f) = f_1 \otimes g_1 + \cdots + f_n \otimes g_n, f_i, g_i \in C[G]$. Since $(v_x \cdot \mathbf{x})(f) = f_1(x)\mathbf{x}(g_1) + \cdots + f_n(x)\mathbf{x}(g_n)$ (cf. (4.4) and Remark 4.2.3), it follows from (4.2) that

$$(v_x \cdot \mathbf{x})(f) = (f * \mathbf{x})(x).$$

Similarly, we have

$$(\mathbf{x} \cdot v_x)(f) = (\mathbf{x} * f)(x).$$

Let us evaluate $v_x \cdot \mathbf{x} \cdot v_{x^{-1}}$. For $1 \leq i \leq n$, let $\mu^*(g_i) = \sum_{j=1}^p u_{ij} \otimes v_{ij}$, where $u_{ij}, v_{ij} \in C[G]$. Then $\rho_{x^{-1}} g_i = \sum_{j=1}^p v_{ij}(x^{-1}) u_{ij}$. Thus $\mathbf{x}(\rho_{x^{-1}} g_i) = \sum_{j=1}^p \mathbf{x}(u_{ij}) v_{ij}(x^{-1})$ and we deduce that

$$\begin{aligned}(v_x \cdot \mathbf{x} \cdot v_{x^{-1}})(f) &= (v_x \otimes ((\mathbf{x} \otimes v_{x^{-1}}) \circ \mu^*) \circ \mu^*)(f) \\ &= \textstyle\sum_{i=1}^n f_i(x) \left(\sum_{j=1}^p \mathbf{x}(u_{ij}) v_{ij}(x^{-1})\right) \\ &= \textstyle\sum_{i=1}^n f_i(x) \mathbf{x}(\rho_{x^{-1}} g_i).\end{aligned}$$

On the other hand, let $i_x : G \to G, y \mapsto xyx^{-1}$. Then

$$(f \circ i_x)(y) = f(xyx^{-1}) = \sum_{i=1}^n f_i(x) g_i(yx^{-1}).$$

Hence

$$f \circ i_x = \sum_{i=1}^n f_i(x)(\rho_{x^{-1}} g_i). \tag{4.9}$$

So we have

$$(v_x \cdot \mathbf{x} \cdot v_{x^{-1}})(f) = \mathbf{x}(f \circ i_x).$$

□

Lemma 4.5.2. *Let H_1, H_2 be algebraic subgroups of G and $\mathfrak{h}_1, \mathfrak{h}_2$ their Lie algebras. If $H_1 \subset H_2$, then $\mathfrak{h}_1 \subset \mathfrak{h}_2$.*

Proof. This is a consequence of Exercise 12 in chapter 2 and Proposition 4.2.4. □

We shall now see that for a fixed element $\mathbf{x} \in \mathfrak{g}$, we can construct a closed subgroup $\mathcal{H}(\mathbf{x})$ which is contained in every closed subgroup H of G such that $\mathbf{x}$ belongs to its Lie algebra $\mathfrak{L}(H)$.

Recall from Remark 4.2.3 that $C[G]^*$ has a structure of associative algebra with unit v_e. If $\omega \in C[G]^*$, we write $\omega^n = \omega \cdot \overset{n}{\ldots} \cdot \omega$, for an integer $n > 0$, $\omega^0 = v_e$.

Lemma 4.5.3. *For $\mathbf{x} \in \mathfrak{g}$, we define $\mathfrak{j}_{\mathbf{x}} := \{f \in C[G] : \mathbf{x}^n(f) = 0 \ \forall n \in \mathbb{N}\}$. Then $\mathfrak{j}_{\mathbf{x}}$ is a prime ideal of $C[G]$.*

Proof. Let us consider the left invariant derivation $\eta(\mathbf{x})$ defined by $\eta(\mathbf{x})(f) = f * \mathbf{x}$, for $f \in C[G]$. By definition of right convolution, we have $\mathbf{x}(f) = (\eta(\mathbf{x})(f))(e)$. We want to prove that

$$\mathbf{x}^n(f) = (\eta^n(\mathbf{x}))(f)(e), \text{ for all } n \in \mathbb{N}. \tag{4.10}$$

We shall proceed by induction on n. If $f \in C[G], \mu^* f = \sum_{i=1}^p f_i \otimes g_i$, then $\eta(\mathbf{x})(f) = \sum_{i=1}^p \mathbf{x}(g_i) f_i$. (See (4.2).) If $n \in \mathbb{N}$, then

$$\eta^{n+1}(\mathbf{x})(f) = \eta^n(\mathbf{x})(\eta(\mathbf{x})(f)) = \eta^n(\mathbf{x})(\sum_{i=1}^p \mathbf{x}(g_i) f_i) = \sum_{i=1}^p \eta^n(\mathbf{x})(f_i)\mathbf{x}(g_i).$$

We have $\mathbf{x}(f) = (\eta(\mathbf{x})(f))(e)$. Now assume that $\mathbf{x}^n(g) = (\eta^n(\mathbf{x}))(g))(e)$, for all $g \in C[G]$. Then

$$\mathbf{x}^{n+1}(f) = (\mathbf{x}^n \otimes \mathbf{x}) \circ \mu^*(f) = \sum_{i=1}^p \mathbf{x}^n(f_i)\mathbf{x}(g_i) = \sum_{i=1}^p (\eta^n(\mathbf{x}))(f_i))(e)\mathbf{x}(g_i).$$

It follows that $\mathbf{x}^{n+1}(f) = (\eta^{n+1}(\mathbf{x}))(f)(e)$.

Now let $n \in \mathbb{N}$ and $f \in \mathfrak{j}_{\mathbf{x}}$. Then

$$\mathbf{x}^n(\eta(\mathbf{x})(f)) = \eta^n(\mathbf{x})(\eta(\mathbf{x})(f))(e) = (\eta^{n+1}(\mathbf{x})(f))(e) = 0.$$

Thus

$$\eta(\mathbf{x})(\mathfrak{j}_{\mathbf{x}}) \subset \mathfrak{j}_{\mathbf{x}}. \tag{4.11}$$

Now let $f \in \mathfrak{j}_{\mathbf{x}}, g \in C[G]$. Then

$$v_e(fg) = f(e)g(e) = 0\,,\ \mathbf{x}(fg) = \mathbf{x}(f)g(e) + f(e)\mathbf{x}(g) = 0.$$

Assume that $\mathbf{x}^n(uv) = 0$ for all $u \in \mathfrak{j}_{\mathbf{x}}$ and $v \in C[G]$. Then

$$\begin{aligned} \mathbf{x}^{n+1}(fg) &= \eta^n(\eta(\mathbf{x})(fg))(e) = \eta^n(\eta(\mathbf{x})(f)g + f\eta(\mathbf{x})(g))(e) \\ &= \mathbf{x}^n(\eta(\mathbf{x})(f))g + \mathbf{x}^n(f\eta(\mathbf{x})(g)) = 0, \end{aligned}$$

since $\eta(\mathbf{x})(f) \in \mathfrak{j}_{\mathbf{x}}$. So $fg \in \mathfrak{j}_{\mathbf{x}}$ and we have proved that $\mathfrak{j}_{\mathbf{x}}$ is an ideal of $C[G]$.

Finally, let $f, g \in C[G] \setminus \mathfrak{j}_{\mathbf{x}}$. Denote by p and q the smallest integers such that $\mathbf{x}^p(f)$ and $\mathbf{x}^q(g)$ are non-zero. Since $\eta(\mathbf{x})$ is a derivation of $C[G]$, we have:

$$\begin{aligned} \mathbf{x}^{p+q}(fg) &= (\eta^{p+q}(\mathbf{x})(fg))(e) \\ &= \left(\sum\nolimits_{m+n=p+q} \frac{(p+q)!}{m!n!} \eta^m(\mathbf{x})(f)\eta^n(\mathbf{x})(g)\right)(e) \\ &= \sum\nolimits_{m+n=p+q} \frac{(p+q)!}{m!n!} \mathbf{x}^m(f)\mathbf{x}^n(g) \\ &= \frac{(p+q)!}{p!q!} \mathbf{x}^p(f)\mathbf{x}^q(g) \neq 0. \end{aligned}$$

Thus $fg \notin \mathfrak{j}_{\mathbf{x}}$. So $\mathfrak{j}_{\mathbf{x}}$ is a prime ideal of $C[G]$. □

We now set

$$\mathcal{H}(\mathbf{x}) := \mathcal{V}(\mathfrak{j}_{\mathbf{x}}).$$

Then $\mathcal{H}(\mathbf{x})$ is an irreducible closed subset of G. Moreover we can prove the following proposition.

Proposition 4.5.4. *Let G be an algebraic group and $\mathbf{x} \in \mathfrak{L}(G)$.*

a) The subvariety $\mathcal{H}(\mathbf{x})$ is a connected and commutative algebraic subgroup of G such that $\mathbf{x} \in \mathfrak{L}(\mathcal{H}(\mathbf{x}))$.

b) Any closed subgroup H of G satisfying $\mathbf{x} \in \mathfrak{L}(H)$ contains $\mathcal{H}(\mathbf{x})$.

Proof. a) We have $\mathcal{H}(\mathbf{x}) = \{x \in G : f(x) = 0, \forall f \in \mathfrak{j}_{\mathbf{x}}\}$. By definition of $\mathfrak{j}(\mathbf{x})$, $e \in \mathcal{H}(\mathbf{x})$. We shall show that if $x \in \mathcal{H}(\mathbf{x})$ and $f \in \mathfrak{j}_{\mathbf{x}}$, then $\lambda_{x^{-1}} f \in \mathfrak{j}_{\mathbf{x}}$. Since the derivation $\eta(\mathbf{x})$ is left invariant (see (4.3)), we have $\eta^n(\mathbf{x})(\lambda_{x^{-1}} f) = \lambda_{x^{-1}}(\eta^n(\mathbf{x})(f))$. From this, we deduce that

$$\mathbf{x}^n(\lambda_{x^{-1}} f) = (\eta^n(\mathbf{x})(\lambda_{x^{-1}} f))(e) = (\lambda_{x^{-1}}(\eta^n(\mathbf{x})(f)))(e) = (\eta^n(\mathbf{x})(f))(x) = 0,$$

as $\eta(\mathbf{x})(\mathfrak{j}_{\mathbf{x}}) \subset \mathfrak{j}_{\mathbf{x}}$, by (4.11). So $\lambda_{x^{-1}} f \in \mathfrak{j}_{\mathbf{x}}$. Now, if $x, y \in \mathcal{H}(\mathbf{x}), f \in \mathfrak{j}_{\mathbf{x}}$, $f(xy) = \lambda_{x^{-1}} f(y) = 0$, since $\lambda_{x^{-1}} f \in \mathfrak{j}_{\mathbf{x}}, y \in G$. So $xy \in \mathcal{H}(\mathbf{x})$.

As $\eta(\mathbf{x})(\mathfrak{j}_{\mathbf{x}}) \subset \mathfrak{j}_{\mathbf{x}}$, we have $\mathbf{x} \in \mathfrak{L}(\mathcal{H}(\mathbf{x}))$ by Lemma 4.2.9.

Let us now prove that $\mathcal{H}(\mathbf{x})$ is commutative. Let $I = \mathcal{I}(\mathcal{H}(\mathbf{x}) \times \mathcal{H}(\mathbf{x})) \subset C[G \times G] \simeq C[G] \otimes C[G]$. We know that $I = \mathfrak{j}_{\mathbf{x}} \otimes C[G] + C[G] \otimes \mathfrak{j}_{\mathbf{x}}$. (See Exercise 7 in chapter 1.) Let $f \in \mathfrak{j}_{\mathbf{x}}$. If $x, y \in \mathcal{H}(\mathbf{x})$, then $\mu^*(f)(x, y) = f(xy) = 0$. Thus $\mu^*(\mathfrak{j}_{\mathbf{x}}) \subset I$.

For $p, q \in \mathbb{N}$, denote by $J_{p,q}$ the kernel of the linear form $\mathbf{x}^p \otimes \mathbf{x}^q$ on $C[G] \otimes C[G]$ and let $J = \bigcap_{p,q} J_{p,q}$. We have $\mathbf{x}^{p+q}(f) = (x^p \otimes x^q)(\mu^*(f))$, so $f \in \mathfrak{j}_{\mathbf{x}}$ implies that $\mu^*(f) \in J$. It follows $J \subset I$. Since it is clear that $I \subset J$, we obtain $I = J$.

Denote by $\zeta : C[G] \otimes C[G] \to C[G] \otimes C[G]$ the linear map defined by $\zeta(f \otimes g) = g \otimes f$. If $p, q \in \mathbb{N}$ and $f \in C[G]$, we have

$$((\mathbf{x}^p \otimes \mathbf{x}^q) \circ \zeta \circ \mu^*)(f) = ((\mathbf{x}^q \otimes \mathbf{x}^p) \circ \mu^*)(f) = \mathbf{x}^{q+p}(f) = ((\mathbf{x}^p \otimes \mathbf{x}^q) \circ \mu^*)(f).$$

Thus $(\mathbf{x}^p \otimes \mathbf{x}^q) \circ (\zeta \circ \mu^* - \mu^*) = 0$. By the preceding, we deduce that $(\zeta \circ \mu^* - \mu^*)(C[G]) \subset I$. If $x, y \in \mathcal{H}(\mathbf{x})$ and $f \in C[G]$, then

$$\begin{aligned} f(xy) - f(yx) &= \mu^*(f)(x, y) - \mu^*(f)(y, x) \\ &= \textstyle\sum_{i=1}^p f_i(x) g_i(y) - \sum_{i=1}^p f_i(y) g_i(x) \\ &= (\mu^* - \zeta \circ \mu^*)(f)(x, y) = 0. \end{aligned}$$

So we have $xy = yx$.

b) Let H be a closed subgroup of G satisfying $\mathbf{x} \in \mathfrak{L}(H)$. Let $\mathfrak{a} = \mathcal{I}(H)$. By Lemma 4.2.9, we have $\eta(\mathbf{x})(\mathfrak{a}) \subset \mathfrak{a}$, so $\eta^n(\mathbf{x})(\mathfrak{a}) \subset \mathfrak{a}$ and $\mathbf{x}^n(\mathfrak{a}) = \{0\}$, for $n \in \mathbb{N}$, by (4.10). Thus $\mathfrak{a} \subset \mathfrak{j}_{\mathbf{x}}$ and $\mathcal{H}(\mathbf{x}) \subset H$. □

Corollary 4.5.5. *Let $\mathfrak{h}$ be a Lie subalgebra of $\mathfrak{g}$ and $\mathcal{H}(\mathfrak{h})$ the intersection of all algebraic subgroups of G whose Lie algebra contains $\mathfrak{h}$.*

a) $\mathcal{H}(\mathfrak{h})$ is connected and $\mathfrak{h} \subset \mathfrak{L}(\mathcal{H}(\mathfrak{h}))$.

b) If H is a closed connected subgroup of G, $\mathfrak{h}$ its Lie algebra, we have $\mathcal{H}(\mathfrak{h}) = H$.

Proof. a) It is clear that $\mathcal{H}(\mathfrak{h})$ is an algebraic subgroup of G and since $\mathfrak{L}(\mathcal{H}(\mathfrak{h})) = \mathfrak{L}(\mathcal{H}(\mathfrak{h})^0)$, $\mathcal{H}(\mathfrak{h})$ is connected. By Proposition 4.5.4, we have $\mathcal{H}(\mathbf{x}) \subset \mathcal{H}(\mathfrak{h})$ for all $\mathbf{x} \in \mathfrak{h}$. So it follows that $\mathbf{x} \in \mathfrak{L}(\mathcal{H}(\mathbf{x})) \subset \mathfrak{L}(\mathcal{H}(\mathfrak{h}))$. Hence $\mathfrak{h} \subset \mathfrak{L}(\mathcal{H}(\mathfrak{h}))$.

b) Since $\mathcal{H}(\mathfrak{h}) \subset H$, we have $\mathfrak{L}(\mathcal{H}(\mathfrak{h})) \subset \mathfrak{h}$. By a), we have equality. Thus $\dim \mathcal{H}(\mathfrak{h}) = \dim H = \dim \mathfrak{h}$. Since H is connected, we deduce that $\mathcal{H}(\mathfrak{h}) = H$. □

Proposition 4.5.6. *Let H_1, H_2 be algebraic subgroups of G and $\mathfrak{h}_1, \mathfrak{h}_2$ their Lie algebras.*

a) Assume that H_1 and H_2 are connected. If $\mathfrak{h}_1 \subset \mathfrak{h}_2$, then $H_1 \subset H_2$.

b) We have $\mathfrak{L}(H_1 \cap H_2) = \mathfrak{h}_1 \cap \mathfrak{h}_2$.

Proof. a) If $\mathfrak{h}_1 \subset \mathfrak{h}_2$, then $\mathcal{H}(\mathfrak{h}_1) \subset \mathcal{H}(\mathfrak{h}_2)$. So $H_1 \subset H_2$ by Corollary 4.5.5.

b) We have $\mathfrak{L}(H_1 \cap H_2) \subset \mathfrak{h}_1 \cap \mathfrak{h}_2$ by Lemma 4.5.2. On the other hand, $\mathcal{H}(\mathfrak{h}_1 \cap \mathfrak{h}_2) \subset H_1 \cap H_2$, by definition of $\mathcal{H}(\mathfrak{h})$. (See Corollary 4.5.5.) So $\mathfrak{h}_1 \cap \mathfrak{h}_2 \subset \mathfrak{L}(\mathcal{H}(\mathfrak{h}_1 \cap \mathfrak{h}_2)) \subset \mathfrak{L}(H_1 \cap H_2)$. $\square$

Lemma 4.5.7. *Let $\varphi : G \to \mathrm{GL}(n, C)$ be a morphism of algebraic groups and $\varphi_{ij} \in C[G]$ such that $\varphi(x) = (\varphi_{ij}(x)_{1\leq i,j\leq n})$, for $x \in G$. If $\mathbf{x} \in \mathfrak{L}(G)$, then $d\varphi(\mathbf{x}) = (\mathbf{x}(\varphi_{ij}))_{1\leq i,j\leq n}$.*

Proof. We have $d\varphi(\mathbf{x})(X_{ij}) = \mathbf{x}(X_{ij} \circ \varphi) = \mathbf{x}(\varphi_{ij})$. $\square$

Proposition 4.5.8. *For G an algebraic group, $\mathfrak{g}$ its Lie algebra, the differential of the adjoint representation Ad_G is $\mathrm{ad}_{\mathfrak{g}}$.*

Proof. Let $x, y \in G$, $\mathbf{x}, \mathbf{y} \in \mathfrak{g}$, $f \in C[G]$ and $\mu^*(f) = \sum_{i=1}^n f_i \otimes g_i$, where $f_i, g_i \in C[G]$. We obtain $\rho_x f = \sum_{i=1}^n g_i(x) f_i$. Hence

$$(\mathbf{y} * f)(x) = \mathbf{y}(\rho_x f) = \sum_{i=1}^n \mathbf{y}(f_i)\, g_i(x). \tag{4.12}$$

It follows that

$$\mathbf{x}(\mathbf{y} * f) = \sum_{i=1}^n \mathbf{y}(f_i)\, \mathbf{x}(g_i) = (\mathbf{y} \cdot \mathbf{x})(f). \tag{4.13}$$

If $\mathbf{x} \in \mathfrak{g}$, $d\,\mathrm{Ad}(\mathbf{x}) \in \mathfrak{gl}(\mathfrak{g})$ may be identified with an element $\ell(\mathbf{x})$ of $\mathrm{End}(\mathfrak{g})$ and for $\psi \in \mathrm{End}(\mathfrak{g})^* \subset C[\mathrm{GL}(\mathfrak{g})]$, we have

$$\psi(\ell(\mathbf{x})) = d\,\mathrm{Ad}(\mathbf{x})(\psi).$$

Let $f \in C[\mathrm{GL}(\mathfrak{g})]$ and $\mathbf{y} \in \mathfrak{g}$. Define $\vartheta_{f,\mathbf{y}} \in \mathrm{End}(\mathfrak{g})^*$ as follows: for $\varphi \in \mathrm{End}(\mathfrak{g})$,

$$\vartheta_{f,\mathbf{y}}(\varphi) = (\varphi(\mathbf{y}))(f).$$

Then for $\mathbf{x}, \mathbf{y} \in \mathfrak{g}$ and $f \in C[\mathrm{GL}(\mathfrak{g})]$, we have

$$(d\,\mathrm{Ad}(\mathbf{x})(\mathbf{y}))(f) = (\ell(\mathbf{x})(\mathbf{y})(f) = \vartheta_{f,\mathbf{y}}(\ell(\mathbf{x})) = d\,\mathrm{Ad}(\mathbf{x})(\vartheta_{f,\mathbf{y}}) = \mathbf{x}(\vartheta_{f,\mathbf{y}} \circ \mathrm{Ad}).$$

On the other hand, if $x \in G$, then

$$(\vartheta_{f,\mathbf{y}} \circ \mathrm{Ad})(x) = (\mathrm{Ad}(x)(\mathbf{y}))(f) = (d\,i_x(\mathbf{y})(f) = \mathbf{y}(f \circ i_x).$$

But if $\mu^*(f) = \sum_{i=1}^n f_i \otimes g_i$, we have

$$f \circ i_x = \sum_{i=1}^{n} f_i(x)(\rho_{x^{-1}} g_i).$$

(See (4.9).) It follows by the definition of left convolution

$$\mathbf{y}(f \circ i_x) = \sum_{i=1}^{n} f_i(x)(\mathbf{y} * g_i)(x^{-1}) = (\sum_{i=1}^{n} (f_i(\mathbf{y} * g_i) \circ \iota))(x).$$

We deduce therefore that

$$\vartheta_{f,\mathbf{y}} \circ \mathrm{Ad} = \sum_{i=1}^{n} (f_i(\mathbf{y} * g_i) \circ \iota).$$

In view of Proposition 4.2.5, we obtain that

$$\mathbf{x}(\vartheta_{f,\mathbf{y}} \circ \mathrm{Ad}) = \sum_{i=1}^{n} \mathbf{x}(f_i)(\mathbf{y} * g_i)(e) - \sum_{i=1}^{n} f_i(e)\mathbf{x}(\mathbf{y} * g_i).$$

It follows from (4.12), (4.13) and the definition of $\mathbf{y} * g_i$ that

$$\mathbf{x}(\vartheta_{f,\mathbf{y}} \circ \mathrm{Ad}) = \sum_{i=1}^{n} \mathbf{x}(f_i)\mathbf{y}(g_i) - \sum_{i=1}^{n} f_i(e)(\mathbf{y} \cdot \mathbf{x})g_i.$$

Finally by the definition of $[\mathbf{x}, \mathbf{y}]$ (4.4) (see also Remark 4.2.3) we obtain

$$\begin{aligned}(d\,\mathrm{Ad}(\mathbf{x})(\mathbf{y}))(f) &= \mathbf{x}(\vartheta_{f,\mathbf{y}} \circ \mathrm{Ad}) = (\mathbf{x} \cdot \mathbf{y})(f) - (v_e \cdot \mathbf{y} \cdot \mathbf{x})(f) \\ &= (\mathbf{x} \cdot \mathbf{y})(f) - (\mathbf{y} \cdot \mathbf{x})(f) = [\mathbf{x}, \mathbf{y}](f).\end{aligned}$$

Hence

$$d\,\mathrm{Ad}(\mathbf{x})(\mathbf{y}) = [\mathbf{x}, \mathbf{y}] = \mathrm{ad}(\mathbf{x})(\mathbf{y}).$$

□

Corollary 4.5.9. *If H is a normal subgroup of G, then the Lie algebra $\mathfrak{h}$ of H is an ideal of $\mathfrak{g}$.*

Proof. If H is normal in G, for $f \in \mathcal{I}(H) \subset C[G]$ and $x \in H$, we have $f \circ i_x \in \mathcal{I}(H)$. Hence if $\mathbf{x} \in \mathfrak{h}$, $\mathbf{x}(f \circ i_x) = 0$ by Lemma 4.2.9. Thus $\mathrm{Ad}(x)(\mathbf{x})(f) = 0$, and this implies that $\mathrm{Ad}(x)(\mathbf{x}) \in \mathfrak{h}$, again by Lemma 4.2.9.

Let $(\mathbf{x}_1, \dots, \mathbf{x}_n)$ be a basis of $\mathfrak{g}$ such that $(\mathbf{x}_1, \dots, \mathbf{x}_p)$ is a basis of $\mathfrak{h}$. If $\mathbf{y} \in \mathfrak{g}$, then Lemma 4.5.7 implies that for $1 \leq j \leq n$, we have $\pi_{ij} \in C[G]$ such that

$$\mathrm{Ad}(x)(\mathbf{x}_j) = \sum_{i=1}^{n} \pi_{ij}(x)\,\mathbf{x}_i\,,\ \ (d\,\mathrm{Ad})(\mathbf{y})(\mathbf{x}_j) = \sum_{i=1}^{n} \mathbf{y}(\pi_{ij})\,\mathbf{x}_i.$$

So $\mathrm{Ad}(x)(\mathfrak{h}) \subset \mathfrak{h}$ implies that $\pi_{ij} = 0$ if $1 \le j \le p$ and $p+1 \le i \le n$. Thus $(d\,\mathrm{Ad})(\mathbf{x})(\mathfrak{h}) \subset \mathfrak{h}$ and the result follows from Proposition 4.5.8. □

Lemma 4.5.10. *Let $\varphi : G \to G$ be a morphism of varieties such that $\varphi(e) = e$ and set $\psi : G \to G, x \mapsto \varphi(x)x^{-1}$. Then*

$$d\psi = d\varphi - Id_{\mathfrak{L}(G)}.$$

Proof. Let $\theta : G \to G \times G$ be the morphism $x \mapsto (\varphi(x), x^{-1})$. Then $\psi = \mu \circ \theta$. If $\mathbf{x} \in \mathfrak{L}(G)$, then by Proposition 4.2.5, we have $d\psi(\mathbf{x}) = d\mu(d\varphi(\mathbf{x}), -\mathbf{x}) = d\varphi(\mathbf{x}) - \mathbf{x}$. □

Proposition 4.5.11. *Let H_1, H_2 be subgroups of G. Let F be the closure of $[H_1, H_2]$ in G. (It is a subgroup by Proposition 3.3.2.) Let $\mathfrak{g}, \mathfrak{h}_1, \mathfrak{h}_2, \mathfrak{f}$ be the Lie algebras of G, H_1, H_2, F respectively. Let $x_1 \in H_1, x_2 \in H_2, \mathbf{x}_1 \in \mathfrak{h}_1, \mathbf{x}_2 \in \mathfrak{h}_2$. Then*

$$[\mathbf{x}_1, \mathbf{x}_2],\ \mathrm{Ad}(x_1)(\mathbf{x}_2) - \mathbf{x}_2,\ \mathrm{Ad}(x_2)(\mathbf{x}_1) - \mathbf{x}_1$$

are elements of $\mathfrak{f}$.

Proof. For $x_1 \in H_1$, denote by $\varphi_{x_1} : H_2 \to F$ the map $x_2 \mapsto [x_1, x_2]$. If $\mathbf{x}_2 \in \mathfrak{h}_2$, then $d\varphi_{x_1}(\mathbf{x}_2) = \mathrm{Ad}(x_1)(\mathbf{x}_2) - \mathbf{x}_2$, by Lemma 4.5.10. Thus $\mathrm{Ad}(x_1)(\mathbf{x}_2) - \mathbf{x}_2 \in \mathfrak{f}$. Similarly, we have $\mathrm{Ad}(x_2)(\mathbf{x}_1) - \mathbf{x}_1 \in \mathfrak{f}$.

It follows that for $\mathbf{x}_2 \in \mathfrak{h}_2$, we have a map $\psi_{\mathbf{x}_2} : H_1 \to \mathfrak{f}$ given by $\psi_{\mathbf{x}_2}(x_1) = \mathrm{Ad}(x_1)(\mathbf{x}_2) - \mathbf{x}_2$. If $\mathbf{x}_1 \in \mathfrak{h}_1$, then by Proposition 4.5.8, $d\psi_{\mathbf{x}_2}(\mathbf{x}_1) = [\mathbf{x}_1, \mathbf{x}_2]$; hence $[\mathbf{x}_1, \mathbf{x}_2] \in \mathfrak{f}$. □

Corollary 4.5.12. *Let G be an algebraic group and $\mathfrak{g}$ its Lie algebra. Then $[\mathfrak{g}, \mathfrak{g}] \subset \mathfrak{L}([G, G])$.* □

Proposition 4.5.13. *Let G be an algebraic group, H_1, H_2 connected normal algebraic subgroups of G. By Proposition 4.4.2, $H_3 := [H_1, H_2]$ is a connected algebraic subgroup of G. Let $\mathfrak{h}_i$ be the Lie algebra of H_i, $i = 1, 2, 3$. We have*

$$\mathfrak{h}_3 = [\mathfrak{h}_1, \mathfrak{h}_2].$$

Proof. We have $[\mathfrak{h}_1, \mathfrak{h}_2] \subset \mathfrak{h}_3$ by Proposition 4.5.11.

For $x, y \in G$, let $[x, y] = xyx^{-1}y^{-1}$ be their commutator. The identity

$$[x,y]^{-1} = [x^{-1}, xyx^{-1}]$$

gives $[x, H_2]^{-1} = [x^{-1}, H_2]$.

We now apply Proposition 3.2.3 to the morphisms

$$\varphi_x : H_2 \to H_3, y \mapsto [x,y], \quad \text{for } x \in H_1$$

and obtain that there exists $n \in \mathbb{N}$ and $x_1, \dots, x_n \in H_1$ such that the map

$$\varphi : H_2^n \to H_3, (y_1, \dots, y_n) \mapsto [x_1, y_1] \dots [x_n, y_n]$$

is surjective. So, by Proposition 2.2.38, there exists $y = (y_1, \dots, y_n) \in H_2^n$ such that

$$d_y\varphi : T_y(H_2^n) \to T_{\varphi y} H_3$$

is surjective.

Let $s : H_2^n \to H_2^n$ and $\theta : H_3 \to H_3$ be defined as follows:

$$s(z_1, \dots, z_n) = (y_1 z_1, \dots, y_n z_n), \quad \theta(z) = z(\varphi(y))^{-1}.$$

These are isomorphisms of varieties. For $1 \le k \le n$ and $z \in H_2$, let

$$C_k(z) = [x_1, y_1] \dots [x_{k-1}, y_{k-1}][x_k, y_k z][x_k, y_k]^{-1} \dots [x_1, y_1]^{-1}.$$

Finally define $\gamma : H_2^n \to H_2^n$ and $v : H_2^n \to H_2$ by

$$\gamma(z_1, \dots, z_n) = (C_1(z_1), \dots, C_n(z_n)), \quad v(z_1, \dots, z_n) = z_1 \dots z_n.$$

We can easily check that

$$v \circ \gamma = \theta \circ \varphi \circ s.$$

Let $\varepsilon := (e, \dots, e) \in H_2^n$. Since θ and s are isomorphisms of varieties, the differential $d_\varepsilon(v \circ \gamma) : T_\varepsilon(H_2^n) \to T_e(H_3) = \mathfrak{h}_3$ of $v \circ \gamma$ at the point ε is surjective.

Let $u_k = [x_1, y_1] \dots [x_k, y_k], 1 \le k \le n$. Then we easily obtain that

$$C_k(z) = i_{u_k}(i_{y_k x_k}(z) i_{y_k}(z^{-1})),$$

where $i_x(z) := xzx^{-1}$. If $t = (t_1, \dots, t_n) \in (T_e(H_2))^n = T_\varepsilon(H_2^n)$, then taking into account the definition of Ad, Proposition 4.2.5 implies that

$$d_\varepsilon(v \circ \gamma)(t) = \sum_{k=1}^{n} \mathrm{Ad}(u_k)(\mathrm{Ad}(y_k)(\mathrm{Ad}(x_k)(t_k) - t_k)).$$

As H_2 is normal in G, $\mathfrak{h}_2$ is an ideal of $\mathfrak{g}$ by Corollary 4.5.9; hence we have $[\mathfrak{h}_2, [\mathfrak{h}_1, \mathfrak{h}_2]] \subset [\mathfrak{h}_1, \mathfrak{h}_2]$. As H_2 is connected, $[H_1, H_2]$ is closed. (See Proposition 4.4.2.) Then Proposition 4.5.11 implies that $\mathrm{Ad}(z)([\mathfrak{h}_1, \mathfrak{h}_2]) \subset [\mathfrak{h}_1, \mathfrak{h}_2]$ for all $z \in H_2$. As $u_k, y_k \in H_2$, to obtain the result, it suffices therefore to prove that if $\mathbf{y} \in \mathfrak{h}_2$ and $x \in H_1$, then $\mathrm{Ad}(x)(\mathbf{y}) - \mathbf{y} \in [\mathfrak{h}_1, \mathfrak{h}_2]$. But this again follows from 4.5.11 and the fact that H_1 is connected. □

Corollary 4.5.14. *a) If G is solvable (resp. nilpotent), then $\mathfrak{g}$ is solvable (resp. nilpotent).*

b) Assume that G is connected. If $\mathfrak{g}$ is solvable (resp. nilpotent), then G is solvable (resp. nilpotent).

Proof. We have $\mathfrak{L}(G) = \mathfrak{L}(G^0)$ and if G is solvable (resp. nilpotent), then so is G^0. So we may assume that G is connected. Then Proposition 4.5.13 implies that $\mathfrak{L}(D^n(G)) = \mathcal{D}^n(\mathfrak{g})$ and $\mathfrak{L}(C^n(G)) = \mathcal{C}^n(\mathfrak{g})$. So the result follows. □

4.6. Subgroups of $\mathrm{SL}(2, C)$

In this section, we give the classification of the subgroups of the special linear group $\mathrm{SL}(2, C)$.

Theorem 4.6.1. *Let G be an algebraic subgroup of $\mathrm{SL}(2, C)$. Then one of the following four cases can occur.*

Case 1. G is triangularizable.

Case 2. G is conjugate to a subgroup of

$$D^+ = \left\{ \begin{pmatrix} c & 0 \\ 0 & c^{-1} \end{pmatrix} : c \in C, c \neq 0 \right\} \cup \left\{ \begin{pmatrix} 0 & c \\ -c^{-1} & 0 \end{pmatrix} : c \in C, c \neq 0 \right\}$$

and case 1 does not hold.

Case 3. G is finite and cases 1 and 2 do not hold.

Case 4. $G = \mathrm{SL}(2, C)$.

Proof. Let G^0 be the identity component of G. As any two-dimensional Lie algebra is solvable (see Exercise 4), by applying Corollary 4.5.14, we obtain that either $\dim G = 3$, in which case $G = \mathrm{SL}(2, C)$, or else G^0 is solvable. In the latter case, G^0 is triangularizable by the Lie-Kolchin Theorem 4.4.6. Assume that G^0 is triangular.

If G^0 is not diagonalizable, then G^0 contains a matrix A of the form

$$A = \begin{pmatrix} 1 & a \\ 0 & 1 \end{pmatrix}, \text{ with } a \neq 0,$$

since an algebraic group contains the unipotent and semisimple parts of all its elements. (See Proposition 4.3.4.) Since G^0 is normal in G (see Proposition 3.2.1a)), any matrix in G conjugates A into a triangular matrix. A direct computation shows that only triangular matrices have this property. Thus G itself is triangular. This is case 1.

Assume next that G^0 is diagonal and infinite, so G contains a non-scalar diagonal matrix B. As G^0 is normal in G, any element of G conjugates B into a diagonal matrix. A direct computation shows that any matrix with this property must be contained in D^+. Therefore either G is diagonal, this being included in case 1, or G is contained in D^+, this being case 2.

Finally we observe that if G^0 is finite (and therefore $G^0 = \{e\}$), then G must also be finite (Proposition 3.2.1a)). This is case 3. □

We shall now determine which finite groups appear as subgroups of $\mathrm{SL}(2, \mathbb{C})$. This is a classical result which can be found in the work of Klein, Jordan, Fuchs, and others. It can be obtained by purely algebraic methods (see e.g. [**Kov**]) but we have preferred to keep the geometric flavor of the classical works, which moreover justifies the names used for the groups obtained. The proof consists in three steps: First prove that if G is a finite subgroup of $\mathrm{SL}(2, \mathbb{C})$, the quotient $G/Z(G)$ is isomorphic to a subgroup of the special orthogonal group $\mathrm{SO}(3, \mathbb{R})$. Then determine the finite groups of space rotations and finally deduce the group G corresponding to each of those.

We recall that a *unitary matrix* is a matrix $U \in \mathrm{GL}(n, \mathbb{C})$ satisfying $U\overline{U}^T = I$, a *special unitary matrix* is an unitary matrix with determinant equal to 1. We consider the *special unitary group* defined by

$$\mathrm{SU}(n) := \{U \in \mathrm{GL}(n, \mathbb{C}) \, : \, U\overline{U}^T = I \text{ and } \det U = 1\}.$$

Lemma 4.6.2. *Let G be a finite subgroup of* $\mathrm{SL}(2, \mathbb{C})$*. Then G is isomorphic to a subgroup of the special unitary group* $\mathrm{SU}(2)$*.*

Proof. We shall prove that there exists an hermitian inner product which is invariant by any element of G. Then, taking a basis of $\mathbb{C}^2$ which is orthonormal with respect to this hermitian product, each element in G is represented by an unitary matrix.

Take any hermitian inner product in $\mathbb{C}^2$ and denote $x * y$ the product of $x, y \in \mathbb{C}^2$. Let $M_1, \ldots, M_r$ be the elements of G. We define

$$x \cdot y = \sum_{i=1}^{r} (M_i x) * (M_i y).$$

Then it is easy to check that $\cdot$ is hermitian. Now, for $M \in G$, we have

$$(Mx) \cdot (My) = \sum_{i=1}^{r} (M_i\, Mx) * (M_i\, My) = x \cdot y,$$

as $M_i\, M$ runs through all the elements of G when M_i does. □

We now consider the Hamilton quaternion algebra $\mathbb{H}$ and denote by $\mathbb{U}$ the group of unit quaternions, i.e. quaternions with norm equal to 1. We take as usual an $\mathbb{R}$-basis of $\mathbb{H}$, $(\mathbf{e}, \mathbf{i}, \mathbf{j}, \mathbf{k})$, where $\mathbf{e}$ denotes the unit element, with the product relations $\mathbf{ij} = \mathbf{k}, \mathbf{i}^2 = \mathbf{j}^2 = \mathbf{k}^2 = -\mathbf{e}, \mathbf{ij} = -\mathbf{ji}$. For a quaternion $q = a\mathbf{e} + b\mathbf{i} + c\mathbf{j} + d\mathbf{k}$, the conjugate quaternion is $\overline{q} = a\mathbf{e} - b\mathbf{i} - c\mathbf{j} - d\mathbf{k}$ and the norm is $|q| = q\overline{q} = a^2 + b^2 + c^2 + d^2$.

Lemma 4.6.3.

$$\mathbb{U} \simeq \mathrm{SU}(2).$$

Proof. To $q = a\mathbf{e} + b\mathbf{i} + c\mathbf{j} + d\mathbf{k} \in \mathbb{U}$, we associate the matrix

$$A_q := \begin{pmatrix} u & v \\ -\overline{v} & \overline{u} \end{pmatrix},$$

where $u = a + bi, v = c + di$. Then it is easy to check that A_q is an unitary matrix and that $q \mapsto A_q$ defines an isomorphism from $\mathbb{U}$ to $\mathrm{SU}(2)$. □

We now intend to associate to an element in the group $\mathbb{U}$ of unit quaternions a rotation of the three dimensional space.

First, for a quaternion $q \in \mathbb{H}$, the mapping $x \mapsto qx$ defines a linear map λ_q from $\mathbb{R}^4$ to $\mathbb{R}^4$, if we consider the underlying vector space structure of $\mathbb{H}$. If we set $q = a\mathbf{e} + b\mathbf{i} + c\mathbf{j} + d\mathbf{k}$, the matrix of λ_q in the basis $(\mathbf{e}, \mathbf{i}, \mathbf{j}, \mathbf{k})$ is

$$A = \begin{pmatrix} a & -b & -c & -d \\ b & a & -d & c \\ c & d & a & -b \\ d & -c & b & a \end{pmatrix}.$$

We have $AA^T = (a^2+b^2+c^2+d^2)I$, hence $(\det A)^2 = (a^2+b^2+c^2+d^2)^4$ and, since $\det A$ contains the term a^4, we have $\det A = (a^2+b^2+c^2+d^2)^2$. We now endow $\mathbb{R}^4$ with the euclidean space structure given by the standard scalar product with respect to the basis $(\mathbf{e}, \mathbf{i}, \mathbf{j}, \mathbf{k})$. We have then $\lambda_q \in SO(4)$ for $|q| = 1$. We shall call λ_q the *left rotation* associated to the unit quaternion q. Moreover, $q \mapsto \lambda_q$ defines a group morphism from $\mathbb{U}$ to SO(4). We can define as well a linear map $\rho_q : \mathbb{R}^4 \to \mathbb{R}^4$ induced by $x \mapsto xq^{-1}$. If $|q| = 1$, its matrix in the basis $(\mathbf{e}, \mathbf{i}, \mathbf{j}, \mathbf{k})$ is

$$\begin{pmatrix} a & b & c & d \\ -b & a & -d & c \\ -c & d & a & -b \\ -d & -c & b & a \end{pmatrix}$$

and we obtain analogously $\rho_q \in SO(4)$ when $|q| = 1$ and a group morphism $\rho : \mathbb{U} \to SO(4), q \mapsto \rho_q$. We call ρ_q the *right rotation* associated to the unit quaternion q. We now want to see that the definition of right and left rotation do not depend on the chosen basis. On the euclidean space $\mathbb{R}^4$, we consider basis changes between orthonormal basis.

Lemma 4.6.4. *We consider a second basis* $(\mathbf{u}, \mathbf{v}, \mathbf{w})$ *in* $\langle \mathbf{e} \rangle^T$. *The two basis* $(\mathbf{e}, \mathbf{i}, \mathbf{j}, \mathbf{k})$ *and* $(\mathbf{e}, \mathbf{u}, \mathbf{v}, \mathbf{w})$ *define identical quaternion algebra structures on* $\mathbb{R}^4$ *if and only if the basis change matrix between them belongs to* SO(4).

Proof. We set $\mathbf{u} = a_{11}\mathbf{i} + a_{21}\mathbf{j} + a_{31}\mathbf{k}, \mathbf{v} = a_{12}\mathbf{i} + a_{22}\mathbf{j} + a_{32}\mathbf{k}, \mathbf{w} = a_{13}\mathbf{i} + a_{23}\mathbf{j} + a_{33}\mathbf{k}$, for some $A := (a_{ij})_{1 \le i,j \le 3} \in GL(3, \mathbb{R})$. By calculation, we obtain $\mathbf{u}^2 = -1$ if and only if $|\mathbf{u}| = 1$, and analogously for $\mathbf{v}$ and $\mathbf{w}$; $\mathbf{uv} = -\mathbf{vu}$ if and only if $\mathbf{u} \perp \mathbf{v}$ and analogously for $\mathbf{u}, \mathbf{w}$ and $\mathbf{v}, \mathbf{w}$. Now, assuming A is orthogonal, we obtain $\mathbf{uv} = (\det A)\mathbf{w}$. □

The preceding lemma tells that, if we consider orthonormal basis with fixed orientation in $\mathbb{R}^4$, the quaternion algebra structure is determined by the first basis vector $\mathbf{e}$. We shall refer to this quaternion algebra as the $\mathbf{e}$-algebra.

Lemma 4.6.5. *Let* $(\mathbf{e}, \mathbf{i}, \mathbf{j}, \mathbf{k})$ *and* $(\mathbf{f}, \mathbf{u}, \mathbf{v}, \mathbf{w})$ *be two orthonormal basis of* $\mathbb{R}^4$.

a) If both basis give the same orientation, then a left (resp. right) rotation in the $\mathbf{e}$*-algebra is a left (resp. right) rotation in the* $\mathbf{f}$*-algebra.*

b) If the two basis give different orientations, then a left (resp. right) rotation in the $\mathbf{e}$*-algebra is a right (resp. left) rotation in the* $\mathbf{f}$*-algebra.*

Proof. a) Let $\mathbf{f}^{-1}$ be the inverse of $\mathbf{f}$ in the $\mathbf{e}$-algebra. Then the right rotation $\rho_{\mathbf{f}}$ sends the basis $(\mathbf{f}, \mathbf{u}, \mathbf{v}, \mathbf{w})$ to the basis $(\mathbf{e}, \mathbf{u}\mathbf{f}^{-1}, \mathbf{v}\mathbf{f}^{-1}, \mathbf{w}\mathbf{f}^{-1})$; hence it is an algebra isomorphism from the $\mathbf{f}$-algebra to the $\mathbf{e}$-algebra. We obtain then that the left (resp. right) rotation associated to a unit quaternion q in the $\mathbf{e}$-algebra is the left (resp. right) rotation associated to the unit quaternion $q\mathbf{f}^{-1}$ (resp. $\mathbf{f}^{-1}q$) in the $\mathbf{e}$-algebra.

b) Taking into account a), it is enough to consider the two basis $(\mathbf{e}, \mathbf{i}, \mathbf{j}, \mathbf{k})$ and $(\mathbf{e}, \mathbf{i}, \mathbf{j}, -\mathbf{k})$. We obtain that multiplication between the basis vectors is reversed when going from one algebra to the other; hence, denoting by $*$ the product in the second algebra, we have $a * b = ba$ for $a, b \in \mathbb{R}^4$. We obtain then that left rotation associated to a unit quaternion q in one algebra is right rotation associated to q^{-1} in the other algebra. □

We have then proved that the groups of left and right rotations are well determined and do not depend on the choice of the orthonormal basis giving the quaternion algebra structure.

Lemma 4.6.6. *Every element in* $\mathrm{SO}(4, \mathbb{R})$ *can be expressed as the product of a left rotation and a right rotation.*

Proof. An element $\sigma \in \mathrm{SO}(4, \mathbb{R})$ has a canonical form

$$\begin{pmatrix} \cos\alpha & -\sin\alpha & 0 & 0 \\ \sin\alpha & \cos\alpha & 0 & 0 \\ 0 & 0 & \cos\beta & -\sin\beta \\ 0 & 0 & \sin\beta & \cos\beta \end{pmatrix}.$$

Looking at the matrices obtained for left and right rotations, we see that $\sigma = \lambda_{q_1} \circ \rho_{q_2}$, for $q_i = \cos\theta_i \mathbf{e} + \sin\theta_i \mathbf{i}$, $i = 1, 2$, with $\theta_1 = (\alpha + \beta)/2, \theta_2 = (\beta - \alpha)/2$. □

We now consider the rotation $\lambda_q \circ \rho_q$ in $\mathbb{R}^4$ given by $x \mapsto qxq^{-1}$. It clearly leaves invariant the axis $\mathbb{R}\mathbf{e}$; hence it leaves globally invariant its orthogonal complement $\mathbb{R}\mathbf{i} + \mathbb{R}\mathbf{j} + \mathbb{R}\mathbf{k}$ and by restriction we obtain an element φ_q in $\mathrm{SO}(3)$ and a morphism $\mathbb{U} \to \mathrm{SO}(3)$, $q \mapsto \varphi_q$.

Lemma 4.6.7.

$$\mathbb{U}/\{\pm 1\} \simeq \mathrm{SO}(3).$$

Proof. We shall prove that the morphism $\mathbb{U} \to \mathrm{SO}(3)$, $q \mapsto \varphi_q$ is surjective and its kernel is $\{\pm 1\}$. A rotation in $\mathrm{SO}(3)$ can be extended to a rotation in $\mathrm{SO}(4)$ leaving invariant the first axis. By Lemma 4.6.6, this last rotation can be expressed as $\lambda_{q_1} \circ \rho_{q_2}$ for some unit quaternions q_1, q_2. Now $\lambda_{q_1} \circ \rho_{q_2}$

fixes $\mathbf{e}$ if and only if $q_1 = q_2$. To determine the kernel, we have $\lambda_q \circ \rho_q = I$ if and only if q commutes with $\mathbf{i}, \mathbf{j}$ and $\mathbf{k}$, which gives $q = \pm 1$. $\square$

We shall now determine the finite subgroups of SO(3), i.e. the finite groups of space rotations. To this end we shall use the following lemma.

Lemma 4.6.8 (Burnside's Lemma). *Let G be a finite group acting on a finite set X. Let r be the number of orbits and for each $\sigma \in G$ let X^σ denote the set of elements in X fixed by σ. Then we have*

$$r = \frac{1}{|G|} \sum_{\sigma \in G} |X^\sigma|. \tag{4.14}$$

Proof. We count in two different ways the number N of pairs $(\sigma, x) \in G \times X$ such that $\sigma x = x$. For each $x \in X$, the elements $\sigma \in G$ such that $\sigma x = x$ form the stabilizer $I(x)$, so

$$N = \sum_{x \in X} |I(x)| = \sum_{x \in X} \frac{|G|}{|Gx|} = |G| \sum_{x \in X} \frac{1}{|Gx|},$$

where Gx denotes the orbit of the element x. Now the sum $\sum(1/|Gx|)$ for x running through the elements in one fixed orbit is equal to 1, so $\sum_{x \in X}(1/|Gx|) = r$ and we obtain $N = |G|r$. On the other hand, for each $\sigma \in G$, the elements $x \in X$ such that $\sigma x = x$ form the set X^σ; hence $N = \sum_{\sigma \in G} |X^\sigma|$. $\square$

Proposition 4.6.9. *Let G be a finite subgroup of* $\mathrm{SO}(3, \mathbb{R})$. *Then G is one of the following groups.*

(1) *The cyclic group C_n generated by a rotation with angle $2\pi/n$ around an axis,*

(2) *the group formed by the identity and three axial symmetries around orthogonal axes, which is isomorphic to $C_2 \times C_2$,*

(3) *the group of rotations leaving invariant an n-sided prism with regular base, which is isomorphic to the dihedral group D_n,*

(4) *the group of rotations leaving invariant a regular polyhedron, which is isomorphic to*
   (a) *the alternating group A_4 in the case of the tetrahedron,*
   (b) *the symmetric group S_4 in the case of the cube and the octahedron,*
   (c) *the alternating group A_5 in the case of the dodecahedron and the icosahedron.*

Proof. This proof is due to Klein and is based on applying Burnside's Lemma 4.6.8 to the action of G on the set $X \subset S^2$ of poles of the rotations in G different from the identity. Here S^2 denotes the sphere of radius 1 centered in the origin and the poles of a rotation are the points of S^2 fixed by it, i.e. the points in which its axis cuts S^2. Note that if $x, -x$ are the poles of $\sigma \in G$, then $\tau x, -\tau x$ are the poles of $\tau\sigma\tau^{-1}$ for $\tau \in G$, so G really acts on X. For this action (4.14) gives

$$r = \frac{1}{n}(2(n-1) + |X|),$$

where $n = |G|$, since X^σ has two elements if $\sigma \neq I$ and is equal to X for $\sigma = I$. Let $\{O_1, \dots, O_r\}$ be the orbits of the action considered and let x_i be a representative of the orbit O_i, for each i. We have then $\sigma x_i \neq x_j$ for all $\sigma \in G$ if $i \neq j$. Set $m_i = |O_i|, n_i = |I(x_i)|$, where $I(x_i)$ denotes the stabilizer of x_i. As $|X| = \sum_{i=1}^r m_i$ and $n = m_i n_i$, we obtain

$$2(1 - \frac{1}{n}) = r - \frac{|X|}{n} = r - \sum_{i=1}^{r} \frac{m_i}{n} = r - \sum_{i=1}^{r} \frac{1}{n_i},$$

which can be written as

$$\sum_{i=1}^{r} (1 - \frac{1}{n_i}) = 2(1 - \frac{1}{n}).$$

As $1 < 2(1-1/n) < 2$ and $n_i \geq 2$, since each pole is stabilized by at least one rotation besides identity, we obtain $1 - 1/n_i \geq 1/2$. This implies $2 \leq r \leq 3$. We now discuss the two cases $r = 2$ and $r = 3$.

$r = 2$: Equation $r - \sum(m_i/n) = 2(1-1/n)$ reduces to $m_1 + m_2 = 2$; hence $m_1 = m_2 = 1$. There is then exactly two poles, which means that all elements in G have the same axis. So G is the cyclic group generated by a rotation with angle $2\pi/n$, which is case 1 in the statement.

$r = 3$: Equation $r - \sum(1/n_i) = 2(1 - 1/n)$ reduces to

$$1 + \frac{2}{n} = \frac{1}{n_1} + \frac{1}{n_2} + \frac{1}{n_3}. \tag{4.15}$$

We can assume $n_1 \leq n_2 \leq n_3$. As the left hand side of (4.15) is bigger than 1, n_1 must be equal to 2. Now $(1/n_2) + (1/n_3)$ must be $> 1/2$; hence either $n_2 = 2$ or $n_2 = 3$. If $n_2 = 2$, n_3 can take any integer value $m \geq 2$ and we obtain $1 + (2/n) = (1/2) + (1/2) + (1/m) \Rightarrow n = 2m$. If $n_2 = 3$, $1/n_3$ must be greater than $1/6$; hence $n_3 = 3, 4$ or 5. Summing up, the possible values for (n_1,n_2,n_3) are $(2,2,m)$, with $m \geq 2$, $(2,3,3),(2,3,4),(2,3,5)$. The rest of the proof is the discussion of each of the cases.

$(2,2,2)$ Equation (4.15) gives $n=4$. Hence, from $n=m_in_i$, we obtain $m_1=m_2=m_3=2$. As $I(x_i)$ has order 2, G has three elements of order 2. Set $I(x_i)=\{I,s_i\}$. We have $s_3(x_1)=-x_1$ and $s_3(x_2)=-x_2$ as any other possibility contradicts $s_i(x_i)=x_i$. So the axis of s_2 and s_3 are orthogonal to the axis of s_1 and, by a similar reasoning, to each other. We obtain then case 2 in the statement.

$(2,2,m)$ Equation (4.15) gives $n=2m$. From $n=m_in_i$, we obtain $m_1=m_2=m, m_3=2$. Now, the group $I(x_3)$ is a group of order m consisting in rotations around the same axis; hence it is cyclic of order m, generated by a rotation ρ with angle $2\pi/m$. The orbit of x_1 contains m elements and is equal to $\{x_1,\rho(x_1),\ldots,\rho^{m-1}(x_1)\}$, which is the set of vertices of an m-sided regular polygon. As $I(x_1)$ has order 2, G contains the axial symmetry around the axis Ox_1, so G is the group of rotations leaving invariant an m-sided polygon or, equivalently the regular prism having it as base. Hence G is the dihedral group of order $n=2m$.

$(2,3,3)$ Equation (4.15) gives $n=12$; hence $m_1=6, m_2=m_3=4$. Let $P_1=x_3,P_2,P_3,P_4$ be the four poles in O_3. As $I(x_3)$ has order 3 and fixes P_1, it permutes cyclically P_2,P_3,P_4; hence these are the vertices of an equilateral triangle. As each $I(P_i)$ is conjugate to $I(x_3)$, the same is true for any three points among P_1,P_2,P_3,P_4, so these are the vertices of a regular tetrahedron and G is the group of rotations leaving it invariant. It is easy to see that $O_2=\{-P_1,-P_2,-P_3,-P_4\}$ and that O_1 is the set of poles corresponding to the lines connecting the middle points of the three pairs of opposite edges of the tetrahedron.

$(2,3,4)$ Equation (4.15) gives $n=24$; hence $m_1=12, m_2=8, m_3=6$. The stabilizer $I(x_3)$ is a group of order 4 which is cyclic as it is a subgroup of SO(2). A generator of $I(x_3)$ permutes cyclically four elements among the six poles in O_3; hence these are the vertices of a regular octahedron and G is the group of rotations leaving it invariant. It contains the rotations with angle $\pi/2$ and axis connecting two opposite vertices as well as the rotations with angle $2\pi/3$ and axis connecting the centers of two opposite faces. Each element of G corresponds to a permutation of the 4-element set of the pairs of opposite faces.

The group G is also the group of rotations leaving invariant the regular hexahedron whose vertices are the middle points of the faces of the octahedron.

$(2,3,5)$ Equation (4.15) gives $n = 60$; hence $m_1 = 30, m_2 = 20, m_3 = 12$. The group $I(x_3)$ is cyclic of order 5. As O_3 has 12 elements and two of them are fixed by $I(x_3)$, this group permutes the remaining ten in two distinct cycles of five elements each, which are the vertices of two regular pentagons, mutually symmetric with respect to the origin. Hence O_3 is the set of vertices of an icosahedron and G is the group of rotations fixing it. It contains the rotations with angle $2\pi/5$ and axis connecting two opposite vertices, the rotations with angle $2\pi/3$ and axis connecting the centers of two opposite faces and the rotations with angle π connecting the middle points of two opposite edges. Each element of G corresponds to an even permutation of the 5-element set of regular tetrahedrons whose vertices are in the center of faces of the icosahedron, as shown in Figure 1. The group G is as well the group of rotations leaving invariant the regular dodecahedron whose vertices are the middle points of the faces of the icosahedron.

□

We shall now determine the finite subgroups of $\mathrm{SL}(2,\mathbb{C})$.

Proposition 4.6.10. *A finite subgroup of* $\mathrm{SL}(2,\mathbb{C})$ *is conjugate to one of the following.*

1. *a cyclic group of order n generated by the matrix*

$$A_\omega = \begin{pmatrix} \omega & 0 \\ 0 & \omega^{-1} \end{pmatrix},$$

for ω a primitive nth root of unity.

2. *the quaternion group generated by the matrices*

$$B := \begin{pmatrix} i & 0 \\ 0 & -i \end{pmatrix}, \quad C := \begin{pmatrix} 0 & 1 \\ -1 & 0 \end{pmatrix}.$$

3. *the dihedral group of order $2n$ generated by the matrices A_ω and C.*
4. *a double cover of a regular polyhedron rotation group, which is isomorphic to*

   (a) the tetrahedral group $2A_4$ generated by the matrices B and

$$D = \frac{1}{2}\begin{pmatrix} -1+i & -1+i \\ 1+i & -1-i \end{pmatrix}.$$

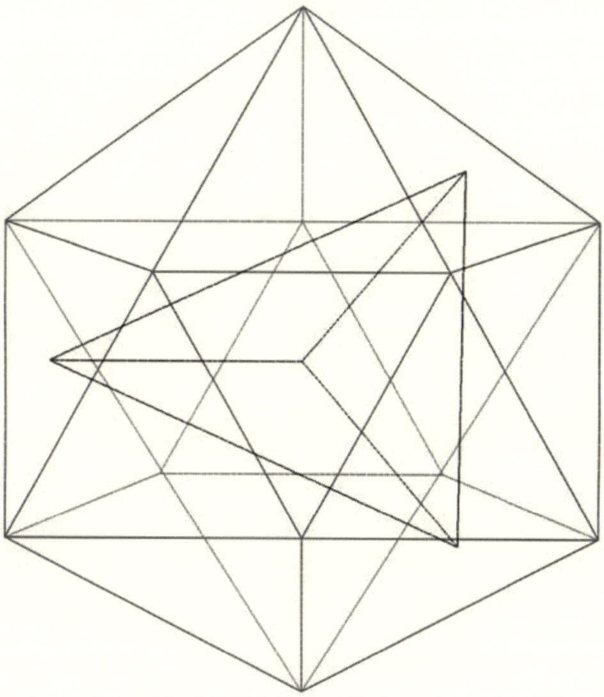

Figure 1. One of the tetrahedrons inscribed in the icosahedron.

(b) the octahedral group $2S_4$ generated by the matrices D and

$$E = \frac{1}{\sqrt{2}} \begin{pmatrix} 1+i & 0 \\ 0 & 1-i \end{pmatrix}.$$

(c) the icosahedral group $2A_5$ generated by B, D and

$$F = \frac{1}{4} \begin{pmatrix} 2i & \beta - i\gamma \\ -\beta - i\gamma & -2i \end{pmatrix},$$

where $\beta = 1 - \sqrt{5}, \gamma = 1 + \sqrt{5}$.

Proof. Cases 1, 2, and 3 follow easily by determining the images of the generators of the subgroups of $\mathrm{SO}(3, \mathbb{R})$ given in cases 1, 2, and 3 of Proposition 4.6.9 by the isomorphisms given in Lemmas 4.6.7 and 4.6.3. We make explicit case 2, i.e. the case of the group formed by the identity and three axial symmetries around orthogonal axes. Taking the symmetry axes to be the coordinate axes, the matrices of the three symmetries are

$$\begin{pmatrix} 1 & 0 & 0 \\ 0 & -1 & 0 \\ 0 & 0 & -1 \end{pmatrix}, \begin{pmatrix} -1 & 0 & 0 \\ 0 & 1 & 0 \\ 0 & 0 & -1 \end{pmatrix}, \begin{pmatrix} -1 & 0 & 0 \\ 0 & -1 & 0 \\ 0 & 0 & 1 \end{pmatrix},$$

which correspond to the quaternions $\mathbf{i}, \mathbf{j}, \mathbf{k}$. These in turn correspond to the matrices in $\mathrm{SL}(2, \mathbb{C})$

$$A_{\mathbf{i}} = \begin{pmatrix} i & 0 \\ 0 & -i \end{pmatrix}, \quad A_{\mathbf{j}} = \begin{pmatrix} 0 & 1 \\ -1 & 0 \end{pmatrix}, \quad A_{\mathbf{k}} = \begin{pmatrix} 0 & i \\ i & 0 \end{pmatrix}.$$

We can assume that $A_{\mathbf{i}}, A_{\mathbf{j}}, A_{\mathbf{k}}$ are sent to $(12)(34), (13)(24), (14)(23)$ respectively by the epimorphism from $\mathrm{SL}(2, \mathbb{C})$ in $\mathrm{SO}(3, \mathbb{R})$ with kernel $\{\pm I\}$.

We now consider the alternating group A_4, generated by the permutations (123) and $(12)(34)$. Conjugation by (123) permutes cyclically $(12)(34)$, $(13)(24)$ and $(14)(23)$. Hence if D is a preimage of (123) in $\mathrm{SL}(2, \mathbb{C})$, it permutes cyclically $A_{\mathbf{i}}, A_{\mathbf{j}}, A_{\mathbf{k}}$ up to multiplication by $-I$. As conjugation by one of the $A's$ on the other two produces multiplication by $-I$, we can look for a matrix D in $\mathrm{SL}(2, \mathbb{C})$ permuting cyclically $A_{\mathbf{i}}, A_{\mathbf{j}}, A_{\mathbf{k}}$ and obtain the matrix D in the statement.

The symmetric group S_4 can be generated by the permutations (123) and (1324). Its preimage in $\mathrm{SL}(2, \mathbb{C})$ is generated by D and a matrix E in $\mathrm{SL}(2, \mathbb{C})$ satisfying $E^2 = A_{\mathbf{i}}$, which is defined up to multiplication by $-I$. We easily find that the matrix E has the form given in the statement.

The alternating group A_5 is generated by the permutations (123) and $(12)(34)$ generating A_4 and the permutation $(12)(45)$. Both $(12)(45)$ and $(123)(12)(45) = (13)(45)$ correspond to rotations with angle π around an axis connecting the middle points of two opposite edges of one of the tetrahedrons inscribed in the icosahedron; hence their preimages in $\mathrm{SL}(2, \mathbb{C})$ have order 4. Now $(12)(34)(12)(45) = (345)$ and we can choose its preimage in $\mathrm{SL}(2, \mathbb{C})$ to have order 3. We look then for a matrix F in $\mathrm{SL}(2, \mathbb{C})$ satisfying $F^2 = (DF)^2 = -I, (A_{\mathbf{i}}F)^3 = I$ and obtain F as in the statement. □

Exercises

(1) Check that the Lie bracket of two derivations is a derivation and that the Lie bracket of two left invariant derivations is again left invariant.

(2) Check that the subspaces of $\mathfrak{gl}(n, C)$ considered in Example 4.1.3 are Lie subalgebras of $\mathfrak{gl}(n, C)$ and that $\mathfrak{sl}(n, C)$ is an ideal of $\mathfrak{gl}(n, C)$.

(3) Write down the proof of Proposition 4.1.7.

(4) Let $\mathfrak{g}$ be a Lie algebra of dimension 2 over the field C. Let (u, v) be a basis of $\mathfrak{g}$ as C-vector space.

a) Prove that the Lie algebra structure of $\mathfrak{g}$ is determined by $[u, v]$.

b) If $[u, v] \neq 0$, prove that the subspace it generates is an ideal of $\mathfrak{g}$, which is independent of the chosen basis and is the unique nontrivial ideal of $\mathfrak{g}$.

c) Prove that $\mathfrak{g}$ is solvable.

(5) Determine the Lie algebra of the orthogonal group

$$\mathrm{O}(n, C) := \{A \in \mathrm{GL}(n, C) : AA^T = Id\}$$

and of the special orthogonal group

$$\mathrm{SO}(n, C) := \{A \in \mathrm{O}(n, C) : \det A = 1\}.$$

(6) Prove the statements in Remark 4.2.3.

(7) Give an example of a linear algebraic group which consists of semisimple elements but is not diagonalizable.

(8) Let V be a finite dimensional C-vector space. If N is a nilpotent endomorphism of V and U an unipotent endomorphism of V, we can define:

$$\exp(N) := \sum_{k=0}^{\infty} \frac{1}{k!} N^k \,,\; \log(U) := \sum_{k=1}^{\infty} \frac{(-1)^{k+1}}{k} (U - Id_V)^k.$$

a) Prove that $\exp(N)$ is an unipotent automorphism of V, $\log(U)$ is a nilpotent endomorphism of V and

$$\exp(\log(U)) = U \,,\; \log(\exp(N)) = N.$$

b) Prove that $\varphi_N : \mathbb{G}_a \to \mathrm{GL}(V), t \mapsto \exp(tN)$ is a morphism of algebraic groups. Deduce that $\varphi_N(\mathbb{G}_a)$ is a connected closed subgroup of $\mathrm{GL}(V)$.

c) For $N \neq 0$, prove that φ_N is an isomorphism of algebraic groups from $\mathbb{G}_a$ onto $\varphi_N(\mathbb{G}_a)$.

(9) Let V be a finite dimensional C-vector space and G be a closed subgroup of $\mathrm{GL}(V)$.
 a) Let U be an unipotent element of G and $N = \log U$. Prove that $\varphi_N(\mathbb{G}_a)$ is the smallest closed subgroup of G containing U.
 b) Let N be a nilpotent endomorphism in $\mathfrak{L}(G)$. Prove that $\mathcal{H}(N)$ is the set of $\exp(tN), t \in C$.

(10) Let V be a vector space of dimension $n > 0$.
 a) Prove that if $x \in \mathrm{End}(V)$ is nilpotent, then $\mathrm{ad}\, x$ is a nilpotent endomorphism of $\mathfrak{gl}(V)$. More precisely, prove that if $x^p = 0$, then $(\mathrm{ad}\, x)^{2p-1} = 0$.
 b) Prove that if $\mathfrak{g}$ is a Lie subalgebra of $\mathfrak{gl}(V)$ consisting of nilpotent endomorphisms, then $V^{\mathfrak{g}} := \{v \in V : \mathbf{x}(v) = 0\,, \forall \mathbf{x} \in \mathfrak{g}\} \neq \{0\}$.

(11) Let G be a connected affine algebraic group of dimension 1. Prove that G is isomorphic either to the multiplicative group $\mathbb{G}_m$ or to the additive group $\mathbb{G}_a$.
Hint: G is commutative by Exercise 9 in chapter 3 and has dimension 1; hence either $G = G_s$ or $G = G_u$.

(12) Prove that the derived subgroup of $\mathrm{GL}(n, C)$ is $\mathrm{SL}(n, C)$ and $\mathrm{SL}(n, C)$ is equal to its own derived subgroup. Conclude that for $n \geq 2$, $\mathrm{GL}(n, C)$ and $\mathrm{SL}(n, C)$ are not solvable.
Hint: Use that each element of one of the subgroups U_{ij} (see the proof of Corollary 3.2.5) is a commutator of elements in $\mathrm{SL}(n, C)$.

(13) a) Prove that an abelian group is nilpotent.
 b) Prove that a nilpotent group is solvable.
 c) Prove that the center of a nontrivial nilpotent group is nontrivial.
 d) If $f : G_1 \to G_2$ is a group morphism, prove that $f(C^i(G_1)) \subset C^i(G_2)$ and that equality holds if f is surjective.

(14) Let G be a group and H a subgroup of G contained in the center of G. Prove that G is nilpotent if and only if G/H is nilpotent.

(15) a) Prove that a group G is nilpotent if and only if there exists a sequence $G = G_1 \supset G_2 \supset \cdots \supset G_{n+1} = \{e\}$ of subgroups of G such that $[G, G_k] \subset G_{k+1}$, for $1 \leq k \leq n$.
 b) If G is a nilpotent group, H a normal subgroup of G, prove that there exists a sequence $H = H_1 \supset H_2 \supset \cdots \supset H_{m+1} = \{e\}$ such that $[G, H_i] \subset H_{i+1}$ for $1 \leq i \leq m$.

(16) Let G be an algebraic group, $A = C[G]$. For B a C-algebra, we consider the set $Hom(A, B)$ of C-algebra morphisms. For $f, g \in Hom(A, B)$, we define its product as the composition

$$A \stackrel{\mu^*}{\to} A \otimes A \stackrel{f \otimes g}{\to} B \otimes B \to B$$

where the last arrow is given by the product in B. This product gives a group structure on $Hom(A, B)$. We denote this group by $G(B)$. The algebra of *dual numbers* is defined by $C[T]/T^2$ or, equivalently, by $C[\delta]$, with $\delta^2 = 0$. We write $\mathcal{T}(G) := G(C[\delta])$.

a) Prove that the points of G are in 1-1 correspondence with $G(C)$.
b) Prove that the projection $C[\delta] \to C, a + b\delta \mapsto a$ induces an epimorphism π from $\mathcal{T}(G)$ onto $G(C)$ with kernel isomorphic to the Lie algebra $\mathfrak{g}$ of the algebraic group G.
c) Prove that the exact sequence

$$0 \to \mathfrak{g} \to \mathcal{T}(G) \xrightarrow{\pi} G(C) \to e$$

splits.

We have $\mathcal{T}(G) = \bigcup_{x \in G} \pi^{-1}(x)$ and $\pi^{-1}(x)$ can be identified with the tangent space $T_x G$ of G at the point x. We call $\mathcal{T}(G)$ the *tangent bundle* of G.

(17) Let G_1, G_2 be affine algebraic groups, $\mathbf{x}_1 \in \mathfrak{L}(G_1), \mathbf{x}_2 \in \mathfrak{L}(G_2)$. For $f_1 \in C[G_1], f_2 \in C[G_2]$, set

$$\theta_{\mathbf{x}_1,\mathbf{x}_2}(f_1 \otimes f_2) = \mathbf{x}_1(f_1) f_2(e_{G_2}) + f_1(e_{G_1}) \mathbf{x}_2(f_2).$$

Prove that the map

$$\begin{array}{rccc} \theta : & \mathfrak{L}(G_1) \times \mathfrak{L}(G_2) & \to & \mathfrak{L}(G_1 \times G_2) \\ & (\mathbf{x}_1, \mathbf{x}_2) & \mapsto & \theta_{\mathbf{x}_1,\mathbf{x}_2} \end{array}$$

is an isomorphism of Lie algebras.

Part 3

Differential Galois Theory

Part 3 is devoted to the Galois Theory of homogeneous linear differential equations referred to as the Picard-Vessiot theory. It parallels classical Galois theory of polynomial equations. The Picard-Vessiot extension of a linear differential equation corresponds to the splitting field of a polynomial and its differential Galois group to the Galois group of the polynomial. We present the fundamental theorem of Picard-Vessiot theory as well as the characterization of homogeneous linear differential equations solvable by quadratures, the analogue of polynomial equations solvable by radicals. The differential Galois group of a linear differential equation defined over a differential field K is a linear algebraic group defined over the constant field of K. The proof of the results mentioned above are based on properties of G-varieties, the existence of quotients for algebraic groups, the decomposition of algebraic groups, the concept of semi-invariant, and the Lie-Kolchin theorem.

In the last chapter, we consider differential equations defined over the field $\mathbb{C}(z)$ of rational functions in one variable over the field $\mathbb{C}$ of complex numbers. We present some classical analytic results concerning local solutions. We end with Kovacic's algorithm, which solves a homogeneous linear differential equation of order 2 by quadratures, whenever it is solvable.

Chapter 5

Picard-Vessiot Extensions

In this chapter we introduce differential rings and differential extensions and define the Picard-Vessiot extension of a homogeneous linear differential equation. We prove its existence and uniqueness in the case when the field of constants of the differential field over which the equation is defined is algebraically closed.

5.1. Derivations

Definition 5.1.1. A *derivation* of a ring A is a map $d : A \to A$ such that

$$d(a+b) = da + db \quad , \quad d(ab) = d(a)\, b + a\, d(b).$$

We write as usual $a' = d(a)$ and $a'', a''', \dots, a^{(n)}$ for successive derivations. By induction, one can prove Leibniz's rule

$$(ab)^{(n)} = a^{(n)}\, b + \cdots + \binom{n}{i} a^{(n-i)} b^{(i)} + \cdots + a\, b^{(n)}.$$

If a' commutes with a, we have $(a^n)' = na^{n-1}a'$. If A has an identity element 1, then necessarily $d(1) = 0$, since $d(1) = d(1.1) = d(1).1 + 1.d(1) \Rightarrow d(1) = 0$. If $a \in A$ is invertible with inverse a^{-1}, we have $a.a^{-1} = 1 \Rightarrow a'a^{-1} + a(a^{-1})' = 0 \Rightarrow (a^{-1})' = -a^{-1}a'a^{-1}$. Hence if a' commutes with a, we get $(a^{-1})' = -a'/a^2$.

Proposition 5.1.2. *If A is an integral domain, a derivation d of A extends to the fraction field $Fr(A)$ in a unique way.*

Proof. For $\frac{a}{b} \in Fr(A)$, we must have $(\frac{a}{b})' = \frac{a'b - ab'}{b^2}$, so there is uniqueness. We extend the derivation to $Fr(A)$ by defining $(\frac{a}{b})' := \frac{a'b - ab'}{b^2}$. If $c \in A \setminus \{0\}$, we have

$$\left(\frac{ac}{bc}\right)' = \frac{(ac)'bc - ac(bc)'}{b^2c^2} = \frac{(a'c + ac')bc - ac(b'c + bc')}{b^2c^2} = \frac{a'b - ab'}{b^2},$$

so the definition is independent of the choice of the representative. Now we have

$$\left(\frac{a}{b} + \frac{c}{d}\right)' = \left(\frac{ad + bc}{bd}\right)' = \frac{(ad + bc)'bd - (ad + bc)(bd)'}{b^2d^2} =$$

$$\frac{(a'd + ad' + b'c + bc')bd - (ad + bc)(b'd + bd')}{b^2d^2} = \frac{a'b - ab'}{b^2} + \frac{c'd - cd'}{d^2},$$

$$\left(\frac{a}{b} \cdot \frac{c}{d}\right)' = \left(\frac{ac}{bd}\right)' = \frac{(ac)'bd - ac(bd)'}{b^2d^2} = \frac{(a'c + ac')bd - ac(b'd + bd')}{b^2d^2} =$$

$$\frac{(a'b - ab')c}{b^2d} + \frac{(c'd - cd')a}{d^2b} = \frac{a'b - ab'}{b^2} \cdot \frac{c}{d} + \frac{a}{b} \cdot \frac{c'd - cd'}{d^2}.$$

□

Remark 5.1.3. If A is a commutative ring with no zero divisors endowed with a derivation and S is a multiplicative system of A, following the same steps as in the proof of Proposition 5.1.2, we can prove that the derivation of A extends to the ring $S^{-1}A$ in a unique way.

5.2. Differential rings

Definition 5.2.1. A *differential ring* is a commutative ring with identity endowed with a derivation. A *differential field* is a differential ring which is a field.

Example 5.2.2. Every commutative ring A with identity can be made into a differential ring with the *trivial derivation* defined by $d(a) = 0, \forall a \in A$.

Over $\mathbb{Z}$ and over $\mathbb{Q}$, the trivial derivation is the only possible one, since $d(1) = 0$, and by induction, $d(n) = d((n-1) + 1) = 0$ and so $d(n/m) = 0$.

Example 5.2.3. The ring of all infinitely differentiable functions on the real line with the usual derivative is a differential ring.

The ring of analytic functions in the complex plane with the usual derivative is a differential ring. In this case, it is an integral domain and so the derivation extends to its fraction field which is the field of meromorphic functions.

Example 5.2.4. Let A be a differential ring and let $A[X]$ be the polynomial ring in one indeterminate over A. A derivation in $A[X]$ extending that of A should satisfy $(\sum a_i X^i)' = \sum (a_i' X^i + a_i i X^{i-1} X')$. We can then extend the derivation of A to $A[X]$ by assigning to X' an arbitrary value in $A[X]$. Analogously, if A is a field, we can extend the derivation of A to the field $A(X)$ of rational functions. By iteration, we can give a differential structure to $A[X_1, \dots, X_n]$ for a differential ring A and to $A(X_1, \dots, X_n)$ for a differential field A.

Example 5.2.5. Let A be a differential ring. We consider the ring $A[X_i]$ of polynomials in the indeterminates $X_i, i \in \mathbb{N} \cup \{0\}$. By defining $X_i' = X_{i+1}$, a unique derivation of $A[X_i]$ is determined. We change notation and write $X = X_0, X^{(n)} = X_n$. We call this procedure the adjunction of a *differential indeterminate* and we use the notation $A\{X\}$ for the resulting differential ring. The elements of $A\{X\}$ are called *differential polynomials* in X. (They are ordinary polynomials in X and its derivatives.) Iterating the process, we define the ring of differential polynomials in n differential indeterminates $X_1, \dots, X_n$ over A by $A\{X_1, \dots, X_n\} = A\{X_1, \dots, X_{n-1}\}\{X_n\}$.

If A is a differential field, then $A\{X_1, \dots, X_n\}$ is a differential integral domain and its derivation extends uniquely to the fraction field. We denote this fraction field by $A\langle X_1, \dots, X_n\rangle$; its elements are *differential rational functions* in $X_1, \dots, X_n$.

Example 5.2.6. If A is a differential ring, we can define a derivation in the ring $M_{n\times n}(A)$ of square $n \times n$ matrices by defining the derivative of a matrix as the matrix obtained by applying the derivation of A to all its entries. Then for $n \geq 2$, $M_{n\times n}(A)$ is a noncommutative ring with derivation.

In any differential ring A, the elements with derivative 0 form a subring called the ring of *constants* and denoted by C_A. If A is a field, so is C_A. The ring of constants C_A contains the image of the ring morphism $\mathbb{Z} \to A$, $1 \mapsto 1$. In the sequel, C_K will denote the constant field of a differential field K.

Definition 5.2.7. Let I be an ideal of a differential ring A. We say that I is a *differential ideal* if $a \in I \Rightarrow a' \in I$, that is, if $d(I) \subset I$.

If I is a differential ideal of the differential ring A, we can define a derivation in the quotient ring A/I by $d(\overline{a}) = \overline{d(a)}$. It is easy to check that this definition does not depend on the choice of the representative in the coset and indeed defines a derivation in A/I.

Definition 5.2.8. If A and B are differential rings, a map $f : A \to B$ is a *differential morphism* if it satisfies

(1) $f(a+b) = f(a) + f(b)$, $f(ab) = f(a)f(b)$, $\forall a, b \in A$; $f(1) = 1$.
(2) $f(a)' = f(a')$, $\forall a \in A$.

If I is a differential ideal, the natural morphism $A \to A/I$ is a differential morphism. The meanings of differential isomorphism and differential automorphism are clear.

Proposition 5.2.9. *If $f : A \to B$ is a differential morphism, then* $\operatorname{Ker} f$ *is a differential ideal and the isomorphism $\overline{f} : A/\operatorname{Ker} f \to \operatorname{Im} f$ is a differential isomorphism.*

Proof. For $a \in \operatorname{Ker} f$, we have $f(a') = f(a)' = 0$, so $a' \in \operatorname{Ker} f$. Hence $\operatorname{Ker} f$ is a differential ideal.

For any $a \in A$, we have $(\overline{f}(\overline{a}))' = (f(a))' = f(a') = \overline{f}(\overline{a'}) = \overline{f}(\overline{a}')$, so $\overline{f}$ is a differential isomorphism. □

5.3. Differential extensions

If A, B are differential rings, A a subring of B, we say that $A \subset B$ is an *extension of differential rings* if the derivation of B restricts to the derivation of A. If S is a subset of B, we denote by $A\{S\}$ the differential A-subalgebra of B generated by S over A, that is, the smallest subring of B containing A, the elements of S and their derivatives. If $K \subset L$ is an extension of differential fields, S a subset of L, we denote by $K\langle S\rangle$ the differential subfield of L generated by S over K. If S is a finite set, we say that the extension $K \subset K\langle S\rangle$ is *differentially finitely generated.*

Proposition 5.3.1. *If K is a differential field, $K \subset L$ a separable algebraic field extension, the derivation of K extends uniquely to L. Moreover, every K-automorphism of L is a differential one.*

Proof. If $K \subset L$ is a finite extension, we have $L = K(\alpha)$, for some α, by the primitive element theorem. If $P(X)$ is the minimal polynomial of α over K, by applying the derivation to $P(\alpha) = 0$, we obtain $P^{(d)}(\alpha) + P'(\alpha)\alpha' = 0$, where $P^{(d)}$ denotes the polynomial obtained from P by deriving its coefficients and P' the derived polynomial. So, $\alpha' = -P^{(d)}(\alpha)/P'(\alpha)$ and the derivation extends uniquely.

Let us now look at the existence. We have $L \simeq K[X]/(P)$. We can extend the derivation of K to $K[X]$ by defining $X' := -P^{(d)}(X)h(X)$ for $h(X) \in K[X]$ such that $h(X)P'(X) \equiv 1 \pmod{P}$. If $h(X)P'(X) = 1 + k(X)P(X)$, we have

$$\begin{aligned} d(P(X)) &= P^{(d)}(X) + P'(X)d(X) \\ &= P^{(d)}(X) + P'(X)(-P^{(d)}(X)h(X)) \\ &= P^{(d)}(X)(1 - P'(X)h(X)) \\ &= -P^{(d)}(X)k(X)P(X). \end{aligned}$$

Therefore (P) is a differential ideal and the quotient field $K[X]/(P)$ is a differential ring.

The general case $K \subset L$ algebraic is obtained from the finite case by applying the Zorn lemma.

Now, if σ is a K-automorphism of L, $\sigma^{-1}d\sigma$ is also a derivation of L extending that of K and by uniqueness, we obtain $\sigma^{-1}d\sigma = d$, and so $d\sigma = \sigma d$, which gives that σ is a differential automorphism. □

5.4. The ring of differential operators

Let K be a differential field with a nontrivial derivation. A *linear differential operator* $\mathcal{L}$ with coefficients in K is a polynomial in the variable D,

$$\mathcal{L} = a_n D^n + a_{n-1} D^{n-1} + \cdots + a_1 D + a_0, \text{ with } a_i \in K.$$

If $a_n \neq 0$, we say that $\mathcal{L}$ has degree n. If $a_n = 1$, we say that $\mathcal{L}$ is monic. The *ring of linear differential operators* with coefficients in K is the noncommutative ring $K[D]$ of polynomials in the variable D with coefficients in K where D satisfies the rule $D\,a = a' + a\,D$ for $a \in K$. We have $\deg(\mathcal{L}_1\mathcal{L}_2) = \deg(\mathcal{L}_1) + \deg(\mathcal{L}_2)$ and then the only left or right invertible elements of $K[D]$ are the elements of $K \setminus \{0\}$. A differential operator acts on K and on differential ring extensions of K with the interpretation $D(y) = d(y)$, for d the derivation in the ring. To the differential operator $\mathcal{L} = a_n D^n + a_{n-1} D^{n-1} + \cdots + a_1 D + a_0$, we associate the linear differential equation

$$\mathcal{L}(Y) = a_n Y^{(n)} + a_{n-1} Y^{(n-1)} + \cdots + a_1 Y' + a_0 Y = 0.$$

Just as for the polynomial ring in one variable over the field K, we have a division algorithm on both left and right.

Lemma 5.4.1. *For $\mathcal{L}_1, \mathcal{L}_2 \in K[D]$ with $\mathcal{L}_2 \neq 0$, there exist unique differential operators Q_l, R_l (resp. Q_r, R_r) in $K[D]$ such that*

$$\begin{aligned} \mathcal{L}_1 = Q_l\mathcal{L}_2 + R_l \quad &\textit{and} \quad \deg R_l < \deg \mathcal{L}_2 \\ (\textit{resp. } \mathcal{L}_1 = \mathcal{L}_2 Q_r + R_r \quad &\textit{and} \quad \deg R_r < \deg \mathcal{L}_2). \end{aligned}$$

The proof of this fact follows the same steps as in the polynomial case.

Corollary 5.4.2. *For each left (resp. right) ideal I of $K[D]$, there exists an element $\mathcal{L} \in K[D]$, unique up to a factor in $K \setminus \{0\}$, such that $I = K[D]\mathcal{L}$ (resp. $I = \mathcal{L}K[D]$).*

Taking into account this corollary, for two linear differential operators $\mathcal{L}_1, \mathcal{L}_2$, the left greatest common divisor will be the unique monic generator of $K[D]\mathcal{L}_1 + K[D]\mathcal{L}_2$ and the left least common multiple will be the unique monic generator of $K[D]\mathcal{L}_1 \cap K[D]\mathcal{L}_2$. Analogously, we can define right GCD and LCM. We can compute left and right GCD with a modified version of Euclid's algorithm.

5.5. Homogeneous linear differential equations

From now on, K will denote a field of **characteristic zero**.

We consider homogeneous linear differential equations over a differential field K, with field of constants C:

$$\mathcal{L}(Y) := Y^{(n)} + a_{n-1}Y^{(n-1)} + \cdots + a_1 Y' + a_0 Y = 0, a_i \in K.$$

If $K \subset L$ is a differential extension, the set of solutions of $\mathcal{L}(Y) = 0$ in L is a C_L-vector space, where C_L denotes the constant field of L. We want to see that its dimension is at most equal to the order n of $\mathcal{L}$.

Definition 5.5.1. Let $y_1, y_2, \dots, y_n$ be elements in a differential field K. The determinant

$$W = W(y_1, y_2, \dots, y_n) := \begin{vmatrix} y_1 & y_2 & \cdots & y_n \\ y_1' & y_2' & \cdots & y_n' \\ \vdots & \vdots & \ddots & \vdots \\ y_1^{(n-1)} & y_2^{(n-1)} & \cdots & y_n^{(n-1)} \end{vmatrix}$$

is the *wrońskian (determinant)* of $y_1, y_2, \dots, y_n$.

Proposition 5.5.2. *Let K be a differential field with field of constants C, and let $y_1, \dots, y_n \in K$. Then $y_1, \dots, y_n$ are linearly independent over C if and only if $W(y_1, \dots, y_n) \neq 0$.*

Proof. Let us assume that $y_1, \dots, y_n$ are linearly dependent over C and let $\sum_{i=1}^n c_i y_i = 0, c_i \in C$ not all zero. By differentiating $n-1$ times this equality, we obtain $\sum_{i=1}^n c_i y_i^{(k)} = 0, k = 0, \dots, n-1$. So the columns of the wrońskian are linearly dependent; hence $W(y_1, \dots, y_n) = 0$.

Reciprocally, let us assume $W(y_1, \dots, y_n) = 0$. We then have n equalities $\sum_{i=1}^n c_i y_i^{(k)} = 0$, $k = 0, \dots, n-1$, with $c_i \in K$ not all zero. We can assume

$c_1 = 1$ and $W(y_2, \ldots, y_n) \neq 0$. By differentiating equality k, we obtain $\sum_{i=1}^n c_i y_i^{(k+1)} + \sum_{i=2}^n c'_i y_i^{(k)} = 0$ and subtracting equality $(k+1)$, we get $\sum_{i=2}^n c'_i y_i^{(k)} = 0, k = 0, \ldots, n-2$. We then obtain a system of homogeneous linear equations in $c'_2, \ldots, c'_n$ with determinant $W(y_2, \ldots, y_n) \neq 0$, so $c'_2 = \cdots = c'_n = 0$, that is, the c_i are constants. □

Taking this proposition into account, we can say "linearly (in)dependent over constants" without ambiguity, since the condition of (non)cancellation of the wrońskian is independent of the field.

Proposition 5.5.3. *Let $\mathcal{L}(Y) = 0$ be a homogeneous linear differential equation of order n over a differential field K. If $y_1, \ldots, y_{n+1}$ are solutions of $\mathcal{L}(Y) = 0$ in a differential extension L of K, then $W(y_1, \ldots, y_{n+1}) = 0$.*

Proof. The last row in the wrońskian is $(y_1^{(n)}, \ldots, y_{n+1}^{(n)})$, which is a linear combination of the preceding ones. □

Corollary 5.5.4. *$\mathcal{L}(Y) = 0$ has at most n solutions in L linearly independent over the field of constants.* □

If $\mathcal{L}(Y) = 0$ is a homogeneous linear differential equation of order n over a differential field K, $y_1, \ldots, y_n$ are n solutions of $\mathcal{L}(Y) = 0$ in a differential extension L of K, linearly independent over the field of constants, we say that $\{y_1, \ldots, y_n\}$ is a *fundamental set of solutions* of $\mathcal{L}(Y) = 0$ in L. Any other solution of $\mathcal{L}(Y) = 0$ in L is a linear combination of $y_1, \ldots, y_n$ with constant coefficients. The next proposition can be proved straightforwardly.

Proposition 5.5.5. *Let $\mathcal{L}(Y) = 0$ be a homogeneous linear differential equation of order n over a differential field K and let $\{y_1, \ldots, y_n\}$ be a basis of the solution space of $\mathcal{L}(Y) = 0$ in a differential extension L of K. Let $z_j = \sum_{i=1}^n c_{ij} y_i$, $j = 1, \ldots, n$, with c_{ij} constants. Then*

$$W(z_1, \ldots, z_n) = \det(c_{ij}) \cdot W(y_1, \ldots, y_n).$$

5.6. The Picard-Vessiot extension

We now define the Picard-Vessiot extension of a homogeneous linear differential equation which is the analogue of the splitting field of a polynomial.

Definition 5.6.1. Given a homogeneous linear differential equation $\mathcal{L}(Y) = 0$ of order n over a differential field K, a differential extension $K \subset L$ is a *Picard-Vessiot extension* for $\mathcal{L}$ if

1. $L = K\langle y_1, \ldots, y_n \rangle$, where $y_1, \ldots, y_n$ is a fundamental set of solutions of $\mathcal{L}(Y) = 0$ in L.

2. Every constant of L lies in K, i.e. $C_L = C_K$.

Remark 5.6.2. Let k be a differential field, $K = k\langle z\rangle$, with $z' = z$, and consider the differential equation $Y' - Y = 0$. As z is a solution to this equation, if we are looking for an analogue of the splitting field, it would be natural to expect that the Picard-Vessiot extension for this equation would be the trivial extension of K. Now, if we adjoin a second differential indeterminate and consider $L = K\langle y\rangle$, with $y' = y$, the extension $K \subset L$ satisfies condition 1 in Definition 5.6.1. But we have $(y/z)' = 0$, so the extension $K \subset L$ adds the new constant y/z. Hence condition 2 in the definition of the Picard-Vessiot extension guarantees its minimality.

In the case when K is a differential field with algebraically closed field of constants C, we shall prove that there exists a Picard-Vessiot extension L of K for a given homogeneous linear differential equation $\mathcal{L}$ defined over K and that it is unique up to differential K-isomorphism.

The idea for the existence proof is to construct a differential K-algebra containing a full set of solutions of the differential equation

$$\mathcal{L}(Y) = Y^{(n)} + a_{n-1}Y^{(n-1)} + \cdots + a_1Y' + a_0Y = 0$$

and then to make the quotient by a maximal differential ideal to obtain an extension not adding constants.

We consider the polynomial ring in n^2 indeterminates

$$K[Y_{ij}, 0 \le i \le n-1, 1 \le j \le n]$$

and extend the derivation of K to $K[Y_{ij}]$ by defining

$$(5.1)\qquad \begin{array}{l} Y'_{ij} = Y_{i+1,j},\ 0 \le i \le n-2, \\ Y'_{n-1,j} = -a_{n-1}Y_{n-1,j} - \cdots - a_1Y_{1j} - a_0Y_{0j}. \end{array}$$

Note that this definition is correct, as we can obtain the preceding ring by defining the ring $K\{X_1, \dots, X_n\}$ in n differential indeterminates and making the quotient by the differential ideal generated by the elements

$$X_j^{(n)} + a_{n-1}X_j^{(n-1)} + \cdots + a_1X_j' + a_0X_j,\ 1 \le j \le n,$$

that is, the ideal generated by these elements and their derivatives. Let $R := K[Y_{ij}][W^{-1}]$ be the localization of $K[Y_{ij}]$ in the multiplicative system of the powers of $W = \det(Y_{ij})$. The derivation of $K[Y_{ij}]$ extends to R in a unique way. The algebra R is called the *full universal solution algebra* for $\mathcal{L}$.

From the next two propositions we shall obtain that a maximal differential ideal P of the full universal solution algebra R is a prime ideal, hence

R/P is an integral domain, and that the fraction field of R/P has the same field of constants as K.

Proposition 5.6.3. *Let K be a differential field and $K \subset R$ be an extension of differential rings. Let I be a maximal element in the set of proper differential ideals of R. Then I is a prime ideal.*

Proof. By passing to the quotient R/I, we can assume that R has no proper differential ideals. Then we have to prove that R is an integral domain. Let us assume that a, b are nonzero elements in R with $ab = 0$. We claim that $d^k(a)b^{k+1} = 0, \forall k \in \mathbb{N}$. Indeed $ab = 0 \Rightarrow 0 = d(ab) = ad(b) + d(a)b$ and, multiplying this equality by b, we obtain $d(a)b^2 = 0$. Now, if it is true for k, $0 = d(d^k(a)b^{k+1}) = d^{k+1}(a)b^{k+1} + (k+1)d^k(a)b^k d(b)$ and, multiplying by b, we obtain $d^{k+1}(a)b^{k+2} = 0$.

Let J now be the differential ideal generated by a, that is, the ideal generated by a and its derivatives. Let us assume that no power of b is zero. By the claim, all elements in J are then zero divisors. In particular $J \neq R$ and, as J contains the nonzero element a, J is a proper differential ideal of R, which contradicts the hypothesis. Therefore, some power of b must be zero.

As b was an arbitrary zero divisor, we have that every zero divisor in R is nilpotent, in particular $a^n = 0$, for some n. We choose n to be minimal. Then $0 = d(a^n) = na^{n-1}d(a)$. As $K \subset R$, we have $na^{n-1} \neq 0$ and so $d(a)$ is a zero divisor. We have then proved that the derivative of a zero divisor is also a zero divisor and so a and all its derivatives are zero divisors and hence nilpotent. In particular, $J \neq R$, so J would be proper and we obtain a contradiction, proving that R is an integral domain. □

Proposition 5.6.4. *Let K be a differential field, with field of constants C, and let $K \subset R$ be an extension of differential rings, such that R is an integral domain, finitely generated as a K-algebra. Let L be the fraction field of R. We assume that C is algebraically closed and that R has no proper differential ideals. Then, L does not contain new constants, i.e. $C_L = C$.*

Proof. 1. First we prove that the elements in $C_L \setminus C$ cannot be algebraic over K. If $\alpha \in C_L$ is algebraic over K, by Exercise 5, it is algebraic over C, so belongs to C.

2. Next we have $C_L \subset R$. Indeed for any $b \in C_L$, we have $b = f/g$, with $f, g \in R$. We consider the ideal of denominators of b, $J = \{h \in R : hb \in R\}$. We have $h \in J \Rightarrow hb \in R \Rightarrow (hb)' = h'b \in R \Rightarrow h' \in J$. Then J is a differential ideal. By hypothesis, R does not contain proper differential ideals, so $J = R$; hence $b = 1.b \in R$.

3. Here we show that for any $b \in C_L$, there exists an element $c \in C$ such that $b - c$ is not invertible in R. Then the ideal $(b-c)R$ is a differential ideal different from R, and is therefore zero. Thus $b = c \in C$.

We now use some results from algebraic geometry. Let $\overline{K}$ be the algebraic closure of K, $\overline{R} = R \otimes_K \overline{K}$. If the element $b \otimes 1 - c \otimes 1 = (b-c) \otimes 1$ is not a unit in $\overline{R}$, then the element $b - c$ will be a nonunit in R. So we can assume that K is algebraically closed. Let V be the affine algebraic variety with coordinate ring R. Then b defines a K-valued function f over V. By Chevalley's theorem (Theorem 2.2.21), its image $f(V)$ is a constructible set in the affine line $\mathbb{A}^1$ and hence either a finite set of points or the complement of a finite set of points. In the second case, as C is infinite, there exists $c \in C$ such that $f(v) = c$, for some $v \in V$ so that $f - c$ vanishes at v and so $b - c$ belongs to the maximal ideal of v. Hence $b - c$ is a nonunit. If $f(V)$ is finite, it consists of a single point, since R is a domain and therefore V is irreducible. So f is constant and b lies in K, hence in C. □

Theorem 5.6.5. *Let K be a differential field with algebraically closed constant field C. Let $\mathcal{L}(Y) = 0$ be a homogeneous linear differential equation defined over K. Let R be the full universal solution algebra for $\mathcal{L}$ and let P be a maximal differential ideal of R. Then P is a prime ideal and the fraction field L of the integral domain R/P is a Picard-Vessiot extension of K for $\mathcal{L}$.*

Proof. R is differentially generated over K by the solutions of $\mathcal{L}(Y) = 0$ and by the inverse of the wrońskian, so R/P as well. By Proposition 5.6.3, P is prime. As P is a maximal differential ideal, R/P does not have proper differential ideals, so by Proposition 5.6.4, $C_L = C$. Moreover, the wrońskian is invertible in R/P and so in particular is nonzero in L. We then have that L contains a fundamental set of solutions of $\mathcal{L}$ and is differentially generated by it over K. Hence L is a Picard-Vessiot extension of K for $\mathcal{L}$. □

In order to obtain uniqueness of the Picard-Vessiot extension, we first prove a normality property.

Proposition 5.6.6. *Let L_1, L_2 be Picard-Vessiot extensions of K for a homogeneous linear differential equation $\mathcal{L}(Y) = 0$ of order n and let $K \subset L$ be a differential field extension with $C_L = C_K$. We assume that $\sigma_i : L_i \to L$ are differential K-morphisms, $i = 1, 2$. Then $\sigma_1(L_1) = \sigma_2(L_2)$.*

Proof. Let $V_i := \{y \in L_i : \mathcal{L}(y) = 0\}, i = 1, 2$, $V := \{y \in L : \mathcal{L}(y) = 0\}$. Then V_i is a C_K-vector space of dimension n and V is a C_K-vector space of dimension at most n. Since σ_i is a differential morphism, we have

$\sigma_i(V_i) \subset V, i = 1, 2$ and so, $\sigma_1(V_1) = \sigma_2(V_2) = V$. From $L_i = K\langle V_i \rangle, i = 1, 2$, we get $\sigma_1(L_1) = \sigma_2(L_2)$. □

Corollary 5.6.7. *Let $K \subset L \subset M$ be differential fields. Assume that L is a Picard-Vessiot extension of K and that M has the same constant field as K. Then any differential K-automorphism of M sends L onto itself.* □

Corollary 5.6.8. *Let K be a differential field, C its field of constants. Assume C is algebraically closed. If L is an algebraic Picard-Vessiot extension of K, then L is a normal algebraic extension of K.*

Proof. Let M be an algebraic extension of K. Then, by Proposition 5.3.1, M is a differential extension of K and every K-automorphism of M is differential. By Exercise 5, the extension $C \subset C_M$ is algebraic as well and so, $C_M = C$. We can now apply Corollary 5.6.7. □

In the next theorem we establish uniqueness up to K-isomorphism of the Picard-Vessiot extension.

Theorem 5.6.9. *Let K be a differential field with algebraically closed field of constants C. Let $\mathcal{L}(Y) = 0$ be a homogeneous linear differential equation defined over K. Let L_1, L_2 be two Picard-Vessiot extensions of K for $\mathcal{L}(Y) = 0$. Then there exists a differential K-isomorphism from L_1 to L_2.*

Proof. We can assume that L_1 is the Picard-Vessiot extension constructed in Theorem 5.6.5. The idea of the proof is to construct a differential extension $K \subset E$ with $C_E = C$ and differential K-morphisms $L_1 \to E$, $L_2 \to E$ and apply Proposition 5.6.6. We consider the ring $A := (R/P) \otimes_K L_2$, which is a differential ring finitely generated as an L_2-algebra, with the derivation defined by $d(x \otimes y) = dx \otimes y + x \otimes dy$. Let Q be a maximal proper differential ideal of A. Its preimage in R/P by the map $R/P \to A$ defined by $a \mapsto a \otimes 1$ is zero, as R/P does not contain proper differential ideals, and it cannot be equal to R/P, as, in this case, Q would be equal to A. So R/P injects in A/Q by $a \mapsto \overline{a \otimes 1}$, and the map $L_2 \to A/Q$ given by $b \mapsto \overline{1 \otimes b}$ is also injective. Now by Proposition 5.6.3, Q is prime and so A/Q is an integral domain. Let E be its fraction field. Now we can apply Proposition 5.6.4 to the L_2-algebra A/Q and obtain $C_E = C_{L_2} = C_K$. By applying Proposition 5.6.6 to the maps $L_1 \hookrightarrow E$, extension of $R/P \hookrightarrow A/Q$, and $L_2 \hookrightarrow A/Q \hookrightarrow E$ we obtain that there exists a differential K-isomorphism $L_1 \to L_2$. □

We now state together the results obtained in Theorems 5.6.5 and 5.6.9.

Theorem 5.6.10. *Let K be a differential field with algebraically closed field of constants C and let $\mathcal{L}(Y) = Y^{(n)} + a_{n-1}Y^{(n-1)} + \cdots + a_1 Y' + a_0 Y = 0$ be* *defined over K. Then there exists a Picard-Vessiot extension L of K for $\mathcal{L}$ and it is unique up to differential K-isomorphism.*

Remark 5.6.11. We have obtained the existence and uniqueness of the Picard-Vessiot extension for a homogeneous linear differential equation defined over a differential field with an algebraically closed field of constants. Seidenberg [**Se**] gives an example of a differential field F and a homogeneous linear differential equation defined over it such that any extension of F containing a nontrivial solution of the equation adds constants. (See Exercise 25.) For an example in which uniqueness fails, see Exercise 26 or see [**Dy**] for a more detailed account of the situation.

Exercises

(1) Prove Leibniz's rule from the definition of a derivation d.

(2) Provide an example of a ring with derivation containing an element a such that $(a^2)' \neq 2\,a\,a'$.

(3) Provide an example of a differential field K and an inseparable field extension $K \subset L$ such that the derivation of K cannot be extended to L and an example in which the extension of the derivation of K to L is not unique.

(4) Given a differential field K and a separably generated field extension $K \subset L$, prove that the derivation of K can be extended to L, giving L a differential field structure, and that for any algebraic field extension F of K contained in L the differential field structure induced by the one in L is unique.

(5) Let K be a differential field with field of constants C, L a differential ring extension of K. Prove that if a constant $a \in L$ is algebraic and separable over K, then it is algebraic over C.

(6) Provide a proof of Lemma 5.4.1.

(7) Find out what is obtained by applying the Maple instructions `rightdivision`, `leftdivision`, `GCRD`, `LCLM`, `DFactor` to differential operators. Work out some examples. Check that the left and right factors of a differential operator are generally different.

(8) Let A be a differential ring, I a radical differential ideal in A.
  a) For a, b elements in A, prove $ab \in I \Rightarrow ab' \in I$ and $a'b \in I$.
  b) Let S be any subset of A. Let $T := \{x \in A : xS \subset I\}$. Prove that T is a radical differential ideal in A.

(9) Let A be a differential ring.
  For a subset S of A, we denote by $\{S\}$ the smallest radical differential ideal containing S. (Note that the intersection of radical differential ideals is a radical differential ideal.)
  a) Let $a \in A, S \subset A$. Prove $a\{S\} \subset \{aS\}$.
  b) Let S, T be subsets of A. Prove $\{S\}\,\{T\} \subset \{ST\}$.

(10) A *Ritt algebra* is a differential ring which is also an algebra over the field $\mathbb{Q}$ of rational numbers.
  a) Let I be a differential ideal in a Ritt algebra A and let a be an element in A. Prove $a^n \in I \Rightarrow (a')^{2n-1} \in I$.
  Hint: Prove $a^{n-k}(a')^{2k-1} \in I$ by induction on k.

b) Prove that in a Ritt algebra the radical of a differential ideal is a differential ideal.

(11) Let A be a differential ring.

a) Let T be a multiplicatively closed subset of A. Prove that a maximal element in the set of radical differential ideals I of A with $I \cap T = \emptyset$ is a prime ideal.

b) Prove that a radical differential ideal I of A is an intersection of prime differential ideals.

Hint: Given $x \notin I$, consider a maximal radical differential ideal containing I and not containing x.

(12) Let K be a differential field, A a differential K-algebra, $a_1, \dots, a_n$ arbitrary elements of A. Prove that there is a unique morphism of differential K- algebras $\varepsilon : K\{Y_1, \dots, Y_n\} \to A$ such that $\varepsilon(Y_i) = a_i, 1 \leq i \leq n$, where $K\{Y_1, \dots, Y_n\}$ denotes the ring of differential polynomials in n differential indeterminates.

(13) Let $K \subset L$ be a differential field extension. The *differential degree of transcendence* of $K \subset L$, difftransdeg$(L|K)$, is either defined as 0 or the supremum of all integer numbers $n \geq 1$ such that there exists a differential subextension $K \subset F$ with F differentially K-isomorphic to the field $K\langle X_1, \dots X_n\rangle$ of differential rational functions in n indeterminates. If difftransdeg$(L|K) = 0$, we say that L is differentially algebraic over K. Prove that $K \subset L$ is differentially algebraic if and only if for each $a \in L$, there exists a differential polynomial $f \in K\{X\}$ such that $f(a) = 0$.

(14) Let A be a differential ring, I an arbitrary ideal of A. Prove that

$$I^\sharp := \{a \in I : a^{(n)} \in I, \text{ for all } n \geq 1\}$$

is the greatest differential ideal contained in I.

(15) Let B a ring, $B[[T]]$ the ring of power series over B. We consider in $B[[T]]$ the derivation given by

$$\left(\sum_{n\geq 0} b_n T^n\right)' = \sum_{n \geq 1} n b_n T^{n-1}.$$

If A is a differential ring, B a Ritt algebra, $\sigma : A \to B$ a ring morphism, we define the *Taylor morphism* $T_\sigma : A \to B[[T]]$ associated to σ by

$$T_\sigma(a) = \sum_{n\geq 0} \frac{\sigma(a^{(n)})}{n!} T^n.$$

Prove the following properties of the Taylor morphism.

a) T_σ is a morphism of differential rings. Its kernel is $(\operatorname{Ker} \sigma)^\sharp$.

b) T_σ is a morphism of C_A-algebras.
c) If B is a reduced ring, then $(\operatorname{Ker}\sigma)^\sharp$ is a radical ideal.
d) If B is a domain, then $(\operatorname{Ker}\sigma)^\sharp$ is a prime ideal.

(16) Let A be a Ritt algebra. Prove that
a) If I is a proper radical ideal of A, then $I^\sharp$ is a radical ideal.
b) If I is a prime ideal of A, then $I^\sharp$ is a prime ideal.
c) Minimal prime ideals over differential ideals are differential.
d) For all $S \subset A$, we have $\{S\} = \sqrt{[S]}$, where $[S]$ denotes the differential ideal generated by S, i.e. the intersection of all differential ideals of A containing S, and $\{S\}$ was defined in Exercise 9.

(17) Provide a proof of Proposition 5.5.5.

(18) We consider the field of rational functions $\mathbb{C}(z)$ with the usual derivation d/dz and the differential operator $D := z(d/dz)$.
a) Prove the equality of differential operators

$$\frac{d^r}{dz^r} \cdot z = r\frac{d^{r-1}}{dz^{r-1}} + z\frac{d^r}{dz^r}, \quad r \geq 1.$$

Deduce by using induction

$$z^r \frac{d^r}{dz^r} = D(D-1)\dots(D-r+1), \quad r \geq 1. \tag{5.2}$$

b) We consider a differential equation

$$Y^{(n)} + a_1(z)Y^{(n-1)} + \dots + a_{n-1}(z)Y' + a_n(z)Y = 0 \tag{5.3}$$

with $a_i(z) \in \mathbb{C}(z)$ and where the derivation is d/dz. Multiplying (5.3) by z^n and using (5.2), we obtain a differential equation in the form

$$D^nY + b_1(z)D^{n-1}Y + \dots + b_{n-1}(z)DY + b_n(z)Y = 0.$$

Give the expression of the rational functions $b_i(z)$ in terms of the $a_i(z)$. Prove that the two following conditions are equivalent
1. $\lim_{z\to 0} z^i a_i(z)$ exists and is finite for all $i = 1, \dots n$.
2. $b_i(z)$ is holomorphic in a neighborhood of 0 for all $i = 1, \dots n$.

(19) *Gauss hypergeometric function* is defined by

$$F(a, b, c; z) = \sum_{n=0}^{\infty} \frac{(a)_n (b)_n}{(c)_n n!} z^n$$

for $a, b, c \in \mathbb{R}, c \notin \mathbb{Z}_{\leq 0}$, $z \in \mathbb{C}$, where the Pochhammer symbol $(x)_n$ is defined by

$$(x)_0 = 1$$
$$(x)_n = x(x+1)\dots(x+n-1).$$

a) Prove that the radius of convergence of the series is 1, except when a or b are nonpositive integers in which case the series is a polynomial.
b) Prove that Gauss hypergeometric function satisfies the differential equation

$$Y'' + \frac{(a+b+1)z - c}{z(z-1)} Y' + \frac{ab}{z(z-1)} Y = 0,$$

where the derivation is d/dz.

(20) *Reduction of order.* If $\mathcal{L}$ is a linear differential operator of degree n with coefficients in a differential field K and $y \in K$ satisfies $\mathcal{L}(y) = 0$, prove that there exists a linear differential operator $\mathcal{L}_1$ of degree $n-1$ with coefficients in K such that

$$\mathcal{L} = \mathcal{L}_1(D - \frac{y'}{y}).$$

Prove that we can obtain a fundamental set of solutions of $\mathcal{L}(Y) = 0$ containing y from a fundamental set of solutions of $\mathcal{L}_1(Y) = 0$.

(21) Let S be a ring, K a subfield of S. Let L_1 and L_2 be K-vector subspaces of S. Prove that the following conditions are equivalent.
1. Whenever $x_1, x_2, \dots, x_n$ are elements of L_1 which are linearly independent over K and $y_1, y_2, \dots, y_m$ are elements of L_2 which are linearly independent over K, then the mn products $x_i y_j$ are also linearly independent over K.
2. Whenever $x_1, x_2, \dots, x_n$ are elements of L_1 which are linearly independent over K, then these elements are also linearly independent over L_2.
3. Whenever $y_1, y_2, \dots, y_m$ are elements of L_2 which are linearly independent over K, then these elements are also linearly independent over L_1.

If these equivalent conditions are satisfied, we say that L_1 and L_2 are *linearly disjoint over* K.

(22) Let S be a differential ring, K a differential subfield of S. Prove that the field K and the subring of constants C_S of S are linearly disjoint over the field of constants C_K of K.

(23) Let K be a differential field with field of constants the field $\mathbb{Q}$ of rational numbers (e.g. $K = \mathbb{Q}(t)$, with derivation given by $t' = 1$). We consider the field $K(Y)$ of rational functions in the variable Y and

extend derivation by $Y' = Y$. Prove that the extension $K(Y)|K(Y^3)$ is a finite differential field extension which is Picard-Vessiot but not normal.

(24) Consider the differential field $K := \mathbb{Q}(t)$, with derivation given by $t' = 1$. Prove that the extension $K(\sqrt[n]{t})|K$, $n > 2$, is a finite differential field extension which is Picard-Vessiot but not normal.

(25) We consider the field $\mathbb{R}$ of real numbers with the trivial derivation and the ring $\mathbb{R}\{X\}$ of differential polynomials in one indeterminate over $\mathbb{R}$. Let A be the differential ring obtained as the quotient of $\mathbb{R}\{X\}$ by the differential ideal generated by $X'' + 4X$. Put α the class of X in A.

a) Prove that the ideal I of A generated by $\alpha'^2 + 4\alpha^2 + 1$ is a prime differential ideal. Let F be the fraction field of A/I with the derivation extended from the one in A/I. Prove that the field of constants of F is $\mathbb{R}$.

b) Let η be a nonzero solution of the differential equation $Y'' + Y = 0$ in a field containing F and let $L := F\langle\eta\rangle$. Prove that the field of constants of L is not $\mathbb{R}$.

Hint: Show that the elements $\gamma_1 := \eta^2 + \eta'^2$ *and* $\gamma_2 := \alpha\eta^2 + \alpha'\eta\eta' - \alpha\eta'^2$ *are constants and that* γ_1 *is not zero. If* $\gamma_1, \gamma_2 \in \mathbb{R}$*, set* $c = \gamma_2/\gamma_1$ *and observe that* $\zeta := \eta'/\eta$ *is a root of the quadratic polynomial* $(c+\alpha)Z^2 - \alpha' Z + c - \alpha$ *whose discriminant is a negative real number.*

(26) We consider the field $\mathbb{R}$ of real numbers with the trivial derivation and the ring $\mathbb{R}\{X\}$ of differential polynomials in one indeterminate over $\mathbb{R}$. Let A be the differential ring obtained as the quotient of $\mathbb{R}\{X\}$ by the differential ideal generated by $X'' + X$. Put α the class of X in A.

Prove that the ideal I_1 (resp. I_2) of A generated by $\alpha'^2 + \alpha^2 - 1$ (resp. $\alpha'^2 + \alpha^2 + 1$) is a prime differential ideal. Let F_1 (resp. F_2) be the fraction field of A/I_1 (resp. A/I_2) with the derivation extended from the one in A/I_1 (resp. A/I_2). Prove that F_1 and F_2 are nonisomorphic Picard-Vessiot extensions for the equation $Y'' + Y = 0$ over $\mathbb{R}$.

(27) A *linear differential equation in matrix form* over a differential field K is an equation of the form

$$Y' = AY, \tag{5.4}$$

where A is an $n \times n$ matrix with entries in K and Y is a vector of length n. A solution to this equation is an element $y \in L^n$, for L a differential field extension of K, satisfying $y' = Ay$, where y' is the vector obtained from y by derivation of each component.

a) Let $y_1, \dots, y_r \in L^n$ be solutions to (5.4). Prove that $y_1, \dots, y_r$ dependent over $L \Rightarrow y_1, \dots, y_r$ dependent over C_L.

b) Prove that the set of solutions to (5.4) in L^n is a C_L-vector space of dimension $\leq n$.

c) A *fundamental matrix* for (5.4) is a matrix $B \in \mathrm{GL}(n, L)$ satisfying $B' = AB$, where B' is the matrix obtained from B by derivation of each entry. Prove that two fundamental matrices for (5.4) differ by a factor in $\mathrm{GL}(n, C_L)$.

d) To a differential operator $\mathcal{L} = D^n + a_{n-1}D^{n-1} + \cdots + a_1 D + a_0 \in K[D]$, we associate the matrix

$$A_{\mathcal{L}} := \begin{pmatrix} 0 & 1 & 0 & \ldots & 0 & 0 \\ 0 & 0 & 1 & \ldots & 0 & 0 \\ & & \ddots & & & \\ & & & \ddots & & \\ 0 & 0 & 0 & \ldots & 0 & 1 \\ -a_0 & -a_1 & -a_2 & \ldots & -a_{n-2} & -a_{n-1} \end{pmatrix}.$$

Write down a fundamental matrix for $Y' = A_{\mathcal{L}} Y$ in terms of a fundamental set of solutions for $\mathcal{L}(Y) = 0$.

(28) Let K be a differential field, $\mathcal{D} := K[D]$ the ring of differential operators over K. A *differential module over K (or $\mathcal{D}$-module)* is a finite dimensional K-vector space which, moreover, is a left $\mathcal{D}$-module. For a differential module $\mathcal{M}$, $e_1, \ldots, e_n$ a K-basis of $\mathcal{M}$, we write

$$De_i = -\sum_j a_{ji} e_j, \text{ where } A = (a_{ij}) \in Mat_{n \times n}(K).$$

a) For $u \in \mathcal{M}$, prove $Du = 0 \Leftrightarrow u' = Au$ (where we identify $\mathcal{M}$ with K^n by means of the chosen basis).

b) A morphism of differential modules is a K-linear map commuting with D. Let $\mathcal{M}_1, \mathcal{M}_2$ be differential modules with bases $(e_1, \ldots, e_n)$, $(v_1, \ldots, v_m)$, respectively. Let $A_1 \in M(n, K), A_2 \in M(m, K)$ be the matrices defining the $\mathcal{D}$-module structures in the chosen bases. Prove that the K-linear map from $\mathcal{M}_1$ to $\mathcal{M}_2$ with matrix $U \in M_{m \times n}(K)$ in the chosen bases is a morphism of differential modules if and only if

$$U' = A_2 U - U A_1.$$

c) Prove that if $m = n$ and U defines an isomorphism from $\mathcal{M}_1$ to $\mathcal{M}_2$, then

$$A_1 = U^{-1} A_2 U - U^{-1} U'.$$

We say that the differential equations $Y' = A_1Y, Y' = A_2Y$ are *equivalent* if $A_1 = U^{-1}A_2U - U^{-1}U'$, for some invertible matrix U with entries in K, i.e. if the associated $\mathcal{D}$-modules are isomorphic.

d) If $\mathcal{M}_1, \mathcal{M}_2$ are differential modules, we can define a differential module structure on $Hom_K(\mathcal{M}_1, \mathcal{M}_2)$ by

$$(D\varphi)(u) = D(\varphi u) - \varphi(Du),$$

for $\varphi \in Hom_K(\mathcal{M}_1, \mathcal{M}_2), u \in \mathcal{M}_1$. Prove that $\varphi \in Hom_K(\mathcal{M}_1, \mathcal{M}_2)$ is a morphism of differential modules if and only if $D\varphi = 0$.

e) Giving to K a differential module structure by $D1 = 0$, we define the dual differential module $\mathcal{M}^*$ of a differential module $\mathcal{M}$ as $Hom_K(\mathcal{M}, K)$ with the differential structure as in d). If A is the matrix giving the differential module structure to $\mathcal{M}$ in a given basis, find the matrix giving the differential module structure to $\mathcal{M}^*$ in the dual basis.

(29) Let $\mathcal{M}$ be a differential module over a differential field K. Let $n = \dim_K \mathcal{M}$. A *cyclic vector* is an element $v \in \mathcal{M}$ such that $(v, Dv, \dots, D^{n-1}v)$ is a K-basis of $\mathcal{M}$.

Assume that the differential field K contains an element x such that $x' = 1$. Let $(e_0, \dots, e_{n-1})$ be a K-basis of $\mathcal{M}$. For $a \in C_K$, we consider the element v_a in $\mathcal{M}$ given by

$$v_a := \sum_{j=0}^{n-1} \frac{(x-a)^j}{j!} \sum_{p=0}^{j} (-1)^p \binom{j}{p} D^p(e_{j-p}).$$

Let us define inductively $c(i,j) \in \mathcal{M}$ by

$$c(0,j) = \begin{cases} \sum_{p=0}^{j} (-1)^p \binom{j}{p} D^p(e_{j-p}) & \text{for } j \leq n-1 \\ 0 & \text{for } j \geq n \end{cases}$$

and

$$c(i+1,j) := D(c(i,j)) + c(i,j+1).$$

a) Prove

$$D^i(v_a) = \sum_{j=0}^{n-1} \frac{(x-a)^j}{j!} c(i,j)$$

and

$$c(i,j) = \sum_{p=0}^{j} (-1)^p \binom{j}{p} D^p(e_{i+j-p}), \quad \text{for } i+j \leq n-1.$$

b) Let T be an indeterminate and let

$$e_i(T) := \sum_{j=0}^{n-1} \frac{T^j}{j!} c(i,j)$$

so that $e_i(x-a) = D^i(v_a)$ and $e_i(0) = e_i$ for all i. We can write the preceding equalities in matrix form

$$\begin{pmatrix} e_0(T) \\ \vdots \\ e_{n-1}(T) \end{pmatrix} = C(T) \begin{pmatrix} e_0 \\ \vdots \\ e_{n-1} \end{pmatrix}$$

for a matrix $C(T)$ whose entries are polynomials in T. Prove that there is at most $n(n-1)$ elements of the form $x-a$ such that $C(x-a)$ is not invertible. Conclude that there exists a set $S \subset C_K$ with at most $n(n-1)$ elements such that, for $a \in C_K \setminus S$, v_a is a cyclic vector.

(30) Let K be a differential field containing at least one nonconstant element.

a) Prove that any differential module over K contains a cyclic vector.

b) Prove that every linear differential equation in matrix form $Y' = AY$ is equivalent to a differential equation $Y' = A_{\mathcal{L}} Y$ associated to some differential operator $\mathcal{L}$.

Hint: For $\mathcal{M}$ *the differential module associated to the system* $Y' = AY$, *use that the dual module* $\mathcal{M}^*$ *has a cyclic vector.*

Chapter 6

The Galois Correspondence

In this chapter we define the differential Galois group of a homogeneous linear differential equation and prove that it has a structure of linear algebraic group over the field of constants of the differential field over which the equation is defined. We establish the fundamental theorem of differential Galois theory, which gives a bijective correspondence between intermediate differential fields of a Picard-Vessiot extension and closed subgroups of its differential Galois group. We characterize differential equations which are solvable by quadratures as those having a differential Galois group with a solvable identity component.

6.1. Differential Galois group

Definition 6.1.1. If $K \subset L$ is a differential field extension, the group $G(L|K)$ of differential K-automorphisms of L is called *differential Galois group* of the extension $K \subset L$. In the case when $K \subset L$ is a Picard-Vessiot extension for the differential equation $\mathcal{L}(Y) = 0$, the group $G(L|K)$ of differential K- automorphisms of L is also referred to as the Galois group of $\mathcal{L}(Y) = 0$ over K. We shall use the notation $\mathrm{Gal}_K(\mathcal{L})$ or $\mathrm{Gal}(\mathcal{L})$ if the base field is clear from the context.

We want to see now that if $K \subset L$ is a Picard-Vessiot extension, then the subfield of L fixed by the action of $G(L|K)$ is equal to K. This fact will be obtained as a corollary of the next proposition. The reader can compare these results with the analogous property of Galois extensions in classical Galois theory.

Proposition 6.1.2. *Let K be a differential field with an algebraically closed field of constants.*

a) If $K \subset L$ is a Picard-Vessiot extension for $\mathcal{L}(Y) = 0$ and $x \in L \setminus K$, then there exists a differential K-automorphism σ of L such that $\sigma(x) \neq x$.

b) Let $K \subset L \subset M$ be extensions of differential fields, where $K \subset L$ and $K \subset M$ are Picard-Vessiot. Then any $\sigma \in G(L|K)$ can be extended to a differential K-automorphism of M.

Proof. a) We can assume that L is the fraction field of R/P with R the full universal solution algebra for $\mathcal{L}$ and P a maximal differential ideal of R. Let $x = a/b$, with $a, b \in R/P$. Then $x \in A := (R/P)[b^{-1}] \subset K$. We consider the differential K-algebra $T := A \otimes_K A \subset L \otimes_K L$. Let $z = x \otimes 1 - 1 \otimes x \in T$. Since $x \notin K$, we have $z \neq 0$, $z' \neq 0$ (if z was a constant, it would be in K) and z is not nilpotent. ($z^n = 0$, for a minimal n would imply $nz^{n-1}z' = 0$; hence $z' = 0$.) We localize T at z and pass to the quotient $T[1/z]/Q$ by a maximal differential ideal Q of $T[1/z]$. Since z is a unit, its image $\overline{z}$ in $T[1/z]/Q$ is nonzero. We have maps $\tau_i : A \to T[1/z]/Q, i = 1, 2$, induced by $w \mapsto w \otimes 1, w \mapsto 1 \otimes w$. The maximality of P implies that R/P has no nontrivial differential ideals, so neither has A; hence the τ_i are injective. Therefore they both extend to differential K-embeddings of L into the fraction field E of $T[1/z]/Q$. By Proposition 5.6.4, E is a no new constants extension of K, so by Proposition 5.6.6, $\tau_1(L) = \tau_2(L)$. On the other hand, $\tau_1(x) - \tau_2(x) = \overline{z} \neq 0$, so $\tau_1(x) \neq \tau_2(x)$. Thus $\tau = \tau_1^{-1}\tau_2$ is a K-differential automorphism of L with $\tau(x) \neq x$.

b) As $L \subset M$ is Picard-Vessiot (for the same differential equation $\mathcal{L}$ as $K \subset M$, seen as defined over L), we can assume that M is the fraction field of R_1/P, where $R_1 = L \otimes_K R$ with R the full universal solution algebra for $\mathcal{L}$ and P a maximal differential ideal of R_1. Then the extension of $\sigma \in G(L|K)$ to M is induced by $\sigma \otimes Id_R$. □

Corollary 6.1.3. *Let K be a differential field with algebraically closed field of constants. If $K \subset L$ is a Picard-Vessiot extension with differential Galois group $G(L|K)$, we have $L^{G(L|K)} = K$, i.e. the subfield of L which is fixed by the action of $G(L|K)$ is equal to K.*

Proof. The inclusion $K \subset L^{G(L|K)}$ is clear from the definition of $G(L|K)$; the inclusion $L^{G(L|K)} \subset K$ is given by Proposition 6.1.2 a). □

Now we see that the differential Galois group of a Picard-Vessiot extension is a linear algebraic group. We first see that the Galois group of a homogeneous linear differential equation of order n defined over the differential field K is isomorphic to a subgroup of the general linear group $\mathrm{GL}(n, C)$ over the constant field C of K. Indeed, if $y_1, y_2, \ldots, y_n$ is a fundamental set of solutions of $\mathcal{L}(Y) = 0$, for each $\sigma \in \mathrm{Gal}(\mathcal{L})$ and for each $j \in \{1, \ldots, n\}$, $\sigma(y_j)$ is again a solution of $\mathcal{L}(Y) = 0$, and so $\sigma(y_j) = \sum_{i=1}^{n} c_{ij} y_i$, for some $c_{ij} \in C_K$. Thus we can associate to each $\sigma \in \mathrm{Gal}(\mathcal{L})$ the matrix $(c_{ij}) \in \mathrm{GL}(n, C)$. Moreover, as $L = K\langle y_1, \dots, y_n \rangle$, a differential K-automorphism of L is determined by the images of the y_j. Hence, we obtain an injective morphism $\mathrm{Gal}(\mathcal{L}) \to \mathrm{GL}(n, C)$ given by $\sigma \mapsto (c_{ij})$. We can then identify $\mathrm{Gal}(\mathcal{L})$ with a subgroup of $\mathrm{GL}(n, C)$, which is determined up to conjugation. Indeed, if we choose a different fundamental set of solutions of $\mathcal{L}(Y) = 0$, the matrix associated to $\sigma \in \mathrm{Gal}(\mathcal{L})$ differs from (c_{ij}) by conjugation by the basis change matrix. We shall see in Proposition 6.2.1 below that $\mathrm{Gal}(\mathcal{L})$ is closed in $\mathrm{GL}(n, C)$ with respect to the Zariski topology (which is defined in Section 1.1). First, we look at some examples.

Example 6.1.4. We consider the differential extension $L = K\langle \alpha \rangle$, with $\alpha' = a \in K$ such that a is not a derivative in K. We say that L is obtained from K by *adjunction of an integral.* We shall prove that α is transcendental over K, $K \subset K\langle \alpha \rangle$ is a Picard-Vessiot extension, and $G(K\langle \alpha \rangle | K)$ is isomorphic to the additive group of $C = C_K$.

Let us assume that α is algebraic over K and write $P(X) = X^n + \sum_{i=1}^{n} b_i X^{n-i}$ its irreducible polynomial over K. Then $0 = P(\alpha) = \alpha^n + \sum_{i=1}^{n} b_i \alpha^{n-i} \Rightarrow 0 = n\alpha^{n-1} a + b_1' \alpha^{n-1} +$ terms of degree $< n-1 \Rightarrow na + b_1' = 0 \Rightarrow a = (-b_1/n)'$ which gives a contradiction.

We now prove that $K\langle \alpha \rangle$ does not contain new constants. Let us assume that the polynomial $\sum_{i=0}^{n} b_i \alpha^{n-i}$, with $b_i \in K$, is constant. Differentiating, we obtain $0 = b_0' \alpha^n + (nb_0 a + b_1')\alpha^{n-1} +$ terms of degree $< n-1 \Rightarrow b_0' = nb_0 a + b_1' = 0 \Rightarrow a = -b_1'/nb_0 = (-b_1/nb_0)'$, contradicting the hypothesis. Let us assume that the rational function $f(\alpha)/g(\alpha)$ is constant, with g monic, of degree ≥ 1, minimal. Differentiating, we obtain

$$0 = \frac{f(\alpha)' g(\alpha) a - f(\alpha) g(\alpha)' a}{g(\alpha)^2} \Rightarrow \frac{f(\alpha)}{g(\alpha)} = \frac{f(\alpha)'}{g(\alpha)'},$$

with $g(\alpha)'$ a nonzero polynomial of lower degree than g, since $g(\alpha)$ is not a constant and g is monic. This is a contradiction.

We observe that 1 and α are solutions of $Y'' - \dfrac{a'}{a} Y' = 0$, linearly independent over the constants, so $K \subset K\langle \alpha \rangle$ is a Picard-Vessiot extension.

A differential K-automorphism of $K\langle \alpha \rangle$ maps α to $\alpha + c$, with $c \in C$ and a mapping $\alpha \mapsto \alpha + c$ induces a differential K-automorphism of $K\langle \alpha \rangle$, for each $c \in C$. So

$$G(K\langle \alpha \rangle | K) \simeq C \simeq \left\{ \begin{pmatrix} 1 & c \\ 0 & 1 \end{pmatrix} \right\} \subset \mathrm{GL}(2, C).$$

Example 6.1.5. We consider the differential extension $L = K\langle\alpha\rangle$, with $\alpha'/\alpha = a \in K \setminus \{0\}$. We say that L is obtained from K by *adjunction of the exponential of an integral.* It is clear that $K\langle\alpha\rangle = K(\alpha)$ and α is a fundamental set of solutions of the differential equation $Y' - aY = 0$. We assume that $C_L = C_K$. We shall prove that if α is algebraic over K, then $\alpha^n \in K$ for some $n \in \mathbb{N}$. The Galois group $G(L|K)$ is isomorphic to the multiplicative group of $C = C_K$ if α is transcendental over K and to a finite cyclic group if α is algebraic over K.

Let us assume that α is algebraic over K and let $P(X) = X^n + a_{n-1}X^{n-1} + \cdots + a_0$ be its irreducible polynomial. Differentiating, we get $0 = P(\alpha)' = P^{(d)}(\alpha) + P'(\alpha)\alpha' = P^{(d)}(\alpha) + P'(\alpha)a\alpha = an\alpha^n + \sum_{k=0}^{n-1}(a'_k + aka_k)\alpha^k$. Then P divides this last polynomial and so $a'_k + aka_k = ana_k \Rightarrow a'_k = a(n-k)a_k, 0 \leq k \leq n-1$. Hence $(\alpha^{n-k}/a_k)' = 0$. In particular, $\alpha^n = ca_0$ for some $c \in C_L = C_K$. Then $P(X)$ divides $X^n - ca_0 \in K[X]$ and so $P(X) = X^n - ca_0$.

For $\sigma \in G(L|K)$, we have $\sigma(\alpha)' = \sigma(\alpha') = \sigma(a\alpha) = a\sigma(\alpha) \Rightarrow (\sigma(\alpha)/\alpha)' = 0 \Rightarrow \sigma(\alpha) = c\alpha$ for some $c \in C_L = C_K$. If α is transcendental over K, for each $c \in C_K$, we can define a differential K-automorphism of L by $\alpha \mapsto c\alpha$. If $\alpha^n = b \in K$, then $\sigma(\alpha)^n = \sigma(\alpha^n) = \sigma(b) = b \Rightarrow c^n = 1 \Rightarrow c$ must be an nth root of unity and $\mathrm{Gal}(L|K)$ is a finite cyclic group.

Example 6.1.6. We consider a differential field K with algebraically closed field of constants C, an irreducible polynomial $P(X) \in K[X]$ of degree n and a splitting field L of $P(X)$ over K. We shall see that $K \subset L$ is a Picard-Vessiot extension. We know by Proposition 5.3.1 that we can extend the derivation in K to L in a unique way by defining for each root x of $P(X)$ in L, $x' = -P^{(d)}(x)h(x)$ for $h(X) \in K[X]$ such that $h(X)P'(X) \equiv 1 (\operatorname{mod} P)$. Moreover by reducing modulo P, we can obtain an expression of x' as a polynomial in x of degree smaller than n. By deriving the expression obtained for x', we obtain an expression for x'' as a polynomial in x which again by reducing modulo P will have degree smaller than n. Iterating the process, we obtain expressions for the successive derivatives of x as polynomials in x of degree smaller than n. Therefore $x, x', \dots, x^{(n)}$ are linearly dependent over K. If we write down this dependence relation, we obtain a homogeneous linear differential equation with coefficients in K satisfied by all the roots of the polynomial P. Now, let us assume that, while computing the successive derivatives of a root x of P, the first dependence relation found gives the differential equation

$$(6.1) \qquad Y^{(k)} + a_{k-1}Y^{(k-1)} + \cdots + a_1 Y' + a_0 Y = 0, a_i \in K, k \leq n.$$

Then, there exist k roots $x_1, \ldots, x_k$ of P with $W(x_1, \ldots, x_k) \neq 0$ since we would otherwise have found a differential equation of order smaller than k satisfied by all the roots of P. Hence L is a Picard-Vessiot extension of K for the equation (6.1) and by Proposition 5.3.1 the differential Galois group of $K \subset L$ coincides with its algebraic Galois group. Note that, as C is algebraically closed, $C_L = C$.

6.2. The differential Galois group as a linear algebraic group

From now on, we assume that **the constant field $C = C_K$ of K is algebraically closed.**

Proposition 6.2.1. *Let K be a differential field with field of constants C, $L = K\langle y_1, \ldots, y_n\rangle$ a Picard-Vessiot extension of K. There exists a set S of polynomials $F(X_{ij}), 1 \leq i,j \leq n$, with coefficients in C such that*

1) If σ is a differential K-automorphism of L and $\sigma(y_j) = \sum_{i=1}^n c_{ij}y_i$, then $F(c_{ij}) = 0, \forall F \in S$.

2) Given a matrix $(c_{ij}) \in \mathrm{GL}(n, C)$ with $F(c_{ij}) = 0, \forall F \in S$, there exists a differential K-automorphism σ of L such that $\sigma(y_j) = \sum_{i=1}^n c_{ij}y_i$.

Proof. Let $K\{Z_1, \ldots, Z_n\}$ be the ring of differential polynomials in n indeterminates over K. We define a differential K-morphism φ from the ring $K\{Z_1, \ldots, Z_n\}$ in L by $Z_j \mapsto y_j$. Then $\Gamma := \operatorname{Ker}\varphi$ is a prime differential ideal of $K\{Z_1, \ldots, Z_n\}$. Let $L[X_{ij}], 1 \leq i,j \leq n$ be the ring of polynomials in the indeterminates X_{ij} with the derivation defined by $X'_{ij} = 0$. We define a differential K-morphism ψ from $K\{Z_1, \ldots, Z_n\}$ to $L[X_{ij}]$ such that $Z_j \mapsto \sum_{i=1}^n X_{ij}y_i$. Let $\Delta := \psi(\Gamma)$. Let $\{w_k\}$ be a basis of the C-vector space L. We write each polynomial in Δ as a linear combination of the w_k with coefficients polynomials in $C[X_{ij}]$. We take S to be the collection of all these coefficients.

1. Let σ be a differential K-automorphism of L and $\sigma(y_j) = \sum_{i=1}^n c_{ij}y_i$. We consider the diagram

$$
\begin{array}{ccccc}
Z_j & \longmapsto & & & y_j \\
 & K\{Z_1,\ldots,Z_n\} & \xrightarrow{\varphi} & L & \\
\big\downarrow & \big\downarrow \psi & & \big\downarrow \sigma & \\
\sum X_{ij} y_i & L[X_{ij}] & \xrightarrow{v} & L & \\
 & X_{ij} & \longmapsto & c_{ij} &
\end{array}
$$

It is clearly commutative. The image of Γ by $\sigma \circ \varphi$ is 0. Its image by $v \circ \psi$ is Δ evaluated in $X_{ij} = c_{ij}$. Therefore all polynomials of Δ vanish at c_{ij}. Writing this down in the basis $\{w_k\}$, we see that all polynomials of S vanish at c_{ij}.

2. Let us now be given a matrix $(c_{ij}) \in \mathrm{GL}(n, C)$ such that $F(c_{ij}) = 0$ for every F in S. We consider the differential morphism

$$
\begin{array}{ccc}
K\{Z_1,\ldots,Z_n\} & \to & K\{y_1,\ldots,y_n\} \\
Z_j & \mapsto & \sum_i c_{ij} y_i
\end{array} .
$$

It is ψ followed by v in the diagram above. By the hypothesis on (c_{ij}), and the definition of the set S, we see that the kernel of this morphism contains Γ and so we have a differential K-morphism

$$
\begin{array}{cccc}
\sigma : & K\{y_1,\ldots,y_n\} & \to & K\{y_1,\ldots,y_n\} \\
 & y_j & \mapsto & \sum_i c_{ij} y_i
\end{array} .
$$

It remains to prove that it is bijective. If u is a nonzero element in the kernel I, then u cannot be algebraic over K, since in this case, the constant term of the irreducible polynomial of u over K would be in I and then I would be the whole ring. But if u is transcendental, we have

$$\operatorname{trdeg}[K\{y_1,\ldots,y_n\} : K] > \operatorname{trdeg}[K\{\sigma y_1,\ldots,\sigma y_n\} : K].$$

On the other hand,

$$\operatorname{trdeg}[K\{y_j, \sigma y_j\} : K] = \operatorname{trdeg}[K\{y_j, c_{ij}\} : K] = \operatorname{trdeg}[K\{y_j\} : K]$$

and analogously we obtain $\operatorname{trdeg}[K\{y_j, \sigma y_j\} : K] = \operatorname{trdeg}[K\{\sigma y_j\} : K]$, which gives a contradiction. Since the matrix (c_{ij}) is invertible, the image contains $y_1, \ldots, y_n$ and so σ is surjective.

Therefore we have that σ is bijective and can be extended to an automorphism

$$\sigma : K\langle y_1, \dots, y_n\rangle \to K\langle y_1, \dots, y_n\rangle.$$

□

This proposition gives that $G(L|K)$ is a closed (in the Zariski topology) subgroup of $\mathrm{GL}(n, C)$ and then a linear algebraic group. (See Section 3.1.)

Remark 6.2.2. The proper closed subgroups of $\mathrm{GL}(1, C) \simeq C^*$ are finite and hence cyclic groups. So for a homogeneous linear differential equation of order 1 the only possible differential Galois groups are C^* or a finite cyclic group, as we saw directly in Example 6.1.5 above.

Remark 6.2.3. In Example 6.1.4 above, the element α is a solution of the nonhomogeneous linear equation $Y' - a = 0$ and we saw that $K \subset K\langle\alpha\rangle$ is a Picard-Vessiot extension for the equation

$$Y'' - \frac{a'}{a}Y' = 0.$$

More generally, we can associate to the equation

$$\mathcal{L}(Y) = Y^{(n)} + a_{n-1}Y^{(n-1)} + \cdots + a_1 Y' + a_0 Y = b,$$

the homogeneous equation $\overline{\mathcal{L}}(Y) = 0$, where

$$\overline{\mathcal{L}} = (d - \frac{b'}{b})\mathcal{L}.$$

It is easy to check that if $y_1, \dots, y_n$ is a fundamental set of solutions of $\mathcal{L}(Y) = 0$ and y_0 is a particular solution of $\mathcal{L}(Y) = b$, then $y_0, y_1, \dots, y_n$ is a fundamental set of solutions of $\overline{\mathcal{L}}(Y) = 0$.

Remark 6.2.4. The full universal solution algebra $R = K[Y_{ij}][W^{-1}]$ constructed before Proposition 5.6.3 is clearly isomorphic, as a K-algebra, to $K \otimes_C C[\mathrm{GL}(n, C)]$, where $C[\mathrm{GL}(n, C)] = C[X_{11}, \dots, X_{nn}, 1/\det]$ denotes the coordinate ring of the algebraic group $\mathrm{GL}(n, C)$. (See Example 3.1.3.) The isomorphism is given by

$$\begin{array}{ccc} K[Y_{ij}][W^{-1}] & \to & K \otimes_C C[\mathrm{GL}(n, C)] \\ Y_{ij} & \mapsto & X_{i+1,j} \qquad 0 \le i \le n-1, 1 \le j \le n. \end{array}$$

If we let $\mathrm{GL}(n, C)$ act on itself by right translations, i.e.

$$\begin{array}{ccc} \mathrm{GL}(n,C)\times \mathrm{GL}(n,C) & \to & \mathrm{GL}(n,C) \\ (g,h) & \mapsto & hg^{-1} \end{array},$$

the corresponding action of $\mathrm{GL}(n,C)$ on $C[\mathrm{GL}(n,C)]$ is

$$\begin{array}{ccc} \mathrm{GL}(n,C)\times C[\mathrm{GL}(n,C)] & \to & C[\mathrm{GL}(n,C)] \\ (g,f) & \mapsto & \rho_g(f): h\mapsto f(hg). \end{array}$$

(See (3.2).) If we take f to be the function X_{ij} sending a matrix in $\mathrm{GL}(n,C)$ to its entry ij, we have $\rho_g(X_{ij})(h) = X_{ij}(hg) = (hg)_{ij} = \sum_{k=1}^n h_{ik}g_{kj}$.

Now $\mathrm{GL}(n,C)$ acts on $K\otimes_C C[\mathrm{GL}(n,C)]$ by acting on the second factor and via the K-algebra isomorphism above we can make $\mathrm{GL}(n,C)$ act on the full universal solution algebra $R = K[Y_{ij}][W^{-1}]$. If P is the maximal differential ideal of R considered in Theorem 5.6.5 and y_{ij} denote the images of the elements Y_{ij} in the quotient R/P, to an element $\sigma\in G = G(L|K)$ such that $\sigma(y_{ij}) = \sum g_{kj}y_{ik}$, we associate the matrix $(g_{ij})\in \mathrm{GL}(n,C)$. Then both actions are compatible and the Galois group $G(L|K)$ can be defined as $\{\sigma\in \mathrm{GL}(n,C) : \sigma(P) = P\}$. So the Galois group $G(L|K)$ is the stabilizer of the C-vector subspace P of R. Using C-bases of P and of $Ann(P) := \{\omega\in Hom(R,C) : \omega(P) = 0\}$, we can write down equations for $G(L|K)$ in $\mathrm{GL}(n,C)$. This gives a second proof that $G(L|K)$ is a closed subgroup of the algebraic group $\mathrm{GL}(n,C)$.

Proposition 6.2.5. *Let K be a differential field with field of constants C. Let $K\subset L$ be a Picard-Vessiot extension with differential Galois group G. Let T be the K-algebra R/P considered in Theorem 5.6.5. We have an isomorphism of $\overline{K}[G]$-modules $\overline{K}\otimes_K T\simeq \overline{K}\otimes_C C[G]$, where $\overline{K}$ denotes the algebraic closure of the field K.*

Proof. We shall use two lemmas. For any field F, we denote by $F[Y_{ij}, 1/\det]$ the polynomial ring in the indeterminates $Y_{ij}, 1\le i,j\le n$ localized with respect to the determinant of the matrix (Y_{ij}). □

Lemma 6.2.6. *Let L be a differential field with field of constants C. We consider $A := L[Y_{ij}, 1/\det]$ and extend the derivation on L to A by setting $Y'_{ij} = 0$. We consider $B := C[Y_{ij}, 1/\det]$ as a subring of A. The map $I\mapsto IA$ from the set of ideals of B to the set of differential ideals of A is a bijection. The inverse map is given by $J\mapsto J\cap B$.*

Proof. Choose a basis $\{v_s\}_{s\in S_1}$ of L over C, including 1. Then $\{v_s\}_{s\in S_1}$ is also a free basis of the B-module A. The differential ideal IA consists of the finite sums $\sum_s \lambda_s v_s$ with all $\lambda_s\in I$. Hence $IA\cap B = I$.

We now prove that any differential ideal J of A is generated by $I = J\cap B$. Let $\{u_s\}_{s\in S}$ be a basis of B over C. Any element $b\in J$ can be written uniquely as a finite sum $\sum_s \mu_s u_s$, with $\mu_s \in L$. By the length $l(b)$ we will mean the number of subindices s with $\mu_s \neq 0$. By induction on the length of b, we shall show that $b \in IA$. When $l(b) = 0, 1$, the result is clear. Assume $l(b) > 1$. We may suppose that $\mu_{s_1} = 1$ for some $s_1 \in S$ and $\mu_{s_2} \in L \setminus C$ for some $s_2 \in S$. Then $b' = \sum_s \mu'_s u_s$ has a length smaller than $l(b)$ and so $b' \in IA$. Similarly $(\mu_{s_2}^{-1} b)' \in IA$. Therefore $(\mu_{s_2}^{-1})' b = (\mu_{s_2}^{-1} b)' - \mu_{s_2}^{-1} b' \in IA$. Since C is the field of constants of L, one has $(\mu_{s_2}^{-1})' \neq 0$ and so $b \in IA$. □

Lemma 6.2.7. *Let K be a differential field with field of constants C. Let $K \subset L$ be a Picard-Vessiot extension with differential Galois group $G(L|K)$. We consider $A := L[Y_{ij}, 1/\det]$, $B := K[Y_{ij}, 1/\det]$. The map $I \mapsto IA$ from the set of ideals of B to the set of $G(L|K)$-stable ideals of A is a bijection. The inverse map is given by $J \mapsto J \cap B$.*

Proof. The proof is similar to that of Lemma 6.2.6. We have to verify that any $G(L|K)$-stable ideal J of A is generated by $I = J \cap B$. Let $\{u_s\}_{s \in S}$ be a basis of B over K. Any element $b \in J$ can be written uniquely as a finite sum $\sum_s \mu_s u_s$, with $\mu_s \in L$. By the length $l(b)$ we will mean the number of subindices s with $\mu_s \neq 0$. By induction on the length of b, we shall show that $b \in IA$. When $l(b) = 0, 1$, the result is clear. Assume $l(b) > 1$. We may suppose that $\mu_{s_1} = 1$ for some $s_1 \in S$. If all $\mu_s \in K$, then $b \in IA$. If not, there exists some $s_2 \in S$ with $\mu_{s_2} \in L \setminus K$. For any $\sigma \in G$, the length of $\sigma b - b$ is less that $l(b)$. Thus $\sigma b - b \in IA$. By Proposition 6.1.2 a), there exists a σ with $\sigma \mu_{s_2} \neq \mu_{s_2}$. As above, one finds $\sigma(\mu_{s_2}^{-1} b) - \mu_{s_2}^{-1} b \in IA$. Then $(\sigma \mu_{s_2}^{-1} - \mu_{s_2}^{-1}) b = \sigma(\mu_{s_2}^{-1} b) - \mu_{s_2}^{-1} b - \sigma(\mu_{s_2}^{-1})(\sigma b - b) \in IA$. As $\sigma \mu_{s_2}^{-1} - \mu_{s_2}^{-1} \in L^*$, it follows that $b \in IA$. □

Proof of Proposition 6.2.5.

We consider the K-algebra $R = K[Y_{ij}, 1/\det]$ with derivation defined by

$$\begin{aligned} Y'_{ij} &= Y_{i+1,j}, \ 0 \leq i \leq n-2, \\ Y'_{n-1,j} &= -a_{n-1} Y_{n-1,j} - \cdots - a_1 Y_{1j} - a_0 Y_{0j} \end{aligned}$$

as in (5.1). We consider as well the L-algebra $L[Y_{ij}, 1/\det]$ with derivation defined by the derivation in L and the preceding formulae. We now consider the C-algebra $C[X_{st}, 1/\det]$ where $X_{st}, 1 \leq s, t \leq n$ are indeterminates, det denotes the determinant of the matrix (X_{st}) and recall that $C[X_{st}, 1/\det]$ is the coordinate algebra $C[\mathrm{GL}(n, C)]$ of the algebraic group $\mathrm{GL}(n, C)$. We consider the action of the group G on $\mathrm{GL}(n, C)$ by translation on the left, i.e.

$$\begin{array}{rcl} G \times \mathrm{GL}(n, C) & \to & \mathrm{GL}(n, C) \\ (g, h) & \mapsto & gh \end{array}$$

which gives the following action of G on $C[\mathrm{GL}(n, C)]$

$$\begin{array}{rcl} G \times C[GL(n, C)] & \to & C[\mathrm{GL}(n, C)] \\ (g, f) & \mapsto & \lambda_g(f) : h \mapsto f(g^{-1}h). \end{array}$$

If we take f to be X_{st}, the action of an element σ of G on X_{st} is multiplication on the left by the inverse of the matrix of σ as an element in $\mathrm{GL}(n, C)$. We consider $C[X_{st}, 1/\det]$ with this G-action and the inclusion $C[X_{st}, 1/\det] \subset L[X_{st}, 1/\det]$. Now we define the relation between the indeterminates Y_{ij} and X_{st} to be given by $(Y_{ij}) = (r_{ab})(X_{st})$, where r_{ab} are the images of the Y_{ab} in the quotient R/P of the ring R by the maximal differential ideal P. We observe that the G-action we have defined on the X_{st} is compatible with the G-action on L if we take the Y_{ij} to be G-invariant. Now, the definition of the derivation for the Y_{ij} and the r_{ab} gives $X'_{st} = 0$. We then have the following rings

$$K[Y_{ij}, \frac{1}{\det}] \subset L[Y_{ij}, \frac{1}{\det}] = L[X_{st}, \frac{1}{\det}] \supset C[X_{st}, \frac{1}{\det}],$$

each of them endowed with a derivation and a G-action which are compatible with each other. Combining Lemmas 6.2.6 and 6.2.7, we obtain a bijection between the set of differential ideals of $K[Y_{ij}, 1/\det]$ and the set of $G(L|K)$-stable ideals of $C[X_{st}, 1/\det]$. A maximal differential ideal of the first ring corresponds to a maximal $G(L|K)$-stable ideal of the second. So, $Q = PL[Y_{ij}, 1/\det] \cap C[X_{st}, 1/\det]$ is a maximal $G(L|K)$-stable ideal of the ring $C[X_{st}, 1/\det]$. By its maximality, Q is a radical ideal and defines a subvariety W of $\mathrm{GL}(n, C)$, which is minimal with respect to $G(L|K)$-invariance. Thus W is a left coset in $\mathrm{GL}(n, C)$ for the group $G(L|K)$ seen as a subgroup of $\mathrm{GL}(n, C)$. Now, by going to the algebraic closure $\overline{K}$ of K, we have an isomorphism from $G_{\overline{K}}$ to $W_{\overline{K}}$ and, correspondingly, an isomorphism $\overline{K} \otimes_C C[G] \simeq \overline{K} \otimes_C C[W]$ between the coordinate rings.

On the other hand, we have ring isomorphisms

$$\begin{array}{rcl} L \otimes_K T & = & L \otimes_K (K[Y_{ij}, \frac{1}{\det}]/P) \\ & \simeq & L[Y_{ij}, \frac{1}{\det}]/(PL[Y_{ij}, \frac{1}{\det}]) \ \simeq \ L \otimes_C (C[X_{st}, \frac{1}{\det}]/Q) \end{array}$$

and so $L \otimes_K T \simeq L \otimes_C C[W]$.

We then have $\overline{L} \otimes_K T \simeq \overline{L} \otimes_C C[W]$, for $\overline{L}$ the algebraic closure of L. This corresponds to an isomorphism of affine varieties $V_{\overline{L}} \simeq W_{\overline{L}}$, where we denote by V the affine subvariety of $\mathrm{GL}(n, K)$ corresponding to the ideal P of $K[Y_{ij}, 1/\det]$. But both W and V are defined over K and so, by Proposition 1.1.29, we obtain $V_{\overline{K}} \simeq W_{\overline{K}}$. Coming back to the corresponding coordinate rings, we get $\overline{K} \otimes_K T \simeq \overline{K} \otimes_C C[W]$. Composing with the isomorphism obtained above, we have $\overline{K} \otimes_K T \simeq \overline{K} \otimes_C C[G]$, as desired. □

Corollary 6.2.8. *Let $K \subset L$ be a Picard-Vessiot extension with differential Galois group $G(L|K)$. We have*

$$\dim G(L|K) = \mathrm{trdeg}[L : K].$$

Proof. The dimension of the algebraic variety G is equal to the Krull dimension of its coordinate ring $C[G]$. (See Section 1.1.) It can be checked that the Krull dimension of a C-algebra remains unchanged when tensoring by a field extension of C. Then Proposition 6.2.5 gives that the Krull dimension of $C[G]$ is equal to the Krull dimension of the algebra T (where T denotes, as in Proposition 6.2.5, the K-algebra R/P considered in Theorem 5.6.5), which by Noether's normalization lemma (Proposition 1.1.8) is equal to the transcendence degree of L over K. □

Remark 6.2.9. Proposition 6.2.5 can be deduced as a corollary from the fact that the maximal spectrum V of the ring T is a G-torsor. (See [**P-S1**], Theorem 1.28.) In fact, the proposition states that $V_{\overline{K}}$ is a trivial $G_{\overline{K}}$-torsor.

6.3. The fundamental theorem of differential Galois theory

The aim of this chapter is to establish the fundamental theorem of Picard-Vessiot theory, which is analogous to the fundamental theorem in classical Galois theory.

If $K \subset L$ is a Picard-Vessiot extension and F an intermediate differential field, i.e. $K \subset F \subset L$, it is clear that $F \subset L$ is a Picard-Vessiot extension (for the same differential equation as $K \subset L$, viewed as defined over F) with differential Galois group $G(L|F) = \{\sigma \in G(L|K) : \sigma_{|F} = Id_F\}$. If H is a subgroup of $G(L|K)$, we denote by L^H the subfield of L fixed by the action of H, i.e. $L^H = \{x \in L : \sigma(x) = x \,, \forall \sigma \in H\}$. Note that L^H is stable under the derivation of L.

Proposition 6.3.1. *Let $K \subset L$ be a Picard-Vessiot extension, $G(L|K)$ its differential Galois group. The correspondences*

$$H \mapsto L^H \quad , \quad F \mapsto G(L|F)$$

define inclusion inverting mutually inverse bijective maps between the set of Zariski closed subgroups H of $G(L|K)$ and the set of differential fields F with $K \subset F \subset L$.

Proof. It is clear that for H_1, H_2 subgroups of $G(L|K)$, we have $H_1 \subset H_2 \Rightarrow L^{H_1} \supset L^{H_2}$ and that for F_1, F_2 intermediate differential fields, $F_1 \subset F_2 \Rightarrow G(L|F_1) \supset G(L|F_2)$.

It is also straightforward to see that, for a subgroup H of G, we have the equality $L^{G(L|L^H)} = L^H$, and, for an intermediate field F, we have $G(L|L^{G(L|F)}) = G(L|F)$.

We have to prove that $L^{G(L|F)} = F$ for each intermediate differential field F of $K \subset L$ and $H = G(L|L^H)$ for each Zariski closed subgroup H of $G(L|K)$. The first equality follows from the fact observed above that $F \subset L$ is a Picard-Vessiot extension and from Corollary 6.1.3. For the second equality, it is clear that if H is a subgroup of $G(L|K)$, the elements in H fix L^H elementwise. We shall now prove that if H is a subgroup (not necessarily closed) of $G = G(L|K)$, then $H' := G(L|L^H)$ is the Zariski closure of H in G. Assume the contrary, i.e. that there exists a polynomial f on $\mathrm{GL}(n, C)$ (where $C = C_K$ and $L|K$ is a Picard-Vessiot extension for an order n differential equation) such that $f_{|H} = 0$ and $f_{|H'} \neq 0$. If $L = K\langle y_1, \dots, y_n\rangle$, we consider the matrices $A = (y_j^{(i)})_{0\leq i\leq n-1, 1\leq j\leq n}$, $B = (u_j^{(i)})_{0\leq i\leq n-1, 1\leq j\leq n}$, where $u_1, \dots, u_n$ are differential indeterminates. We let the Galois group act on the right, i.e we define the matrix M_σ of $\sigma \in G(L|K)$ such that $(\sigma(y_1), \dots, \sigma(y_n)) = (y_1, \dots, y_n)M_\sigma$. We note that, as $W(y_1, \dots, y_n) \neq 0$, the matrix A is invertible and we define the polynomial $F(u_1, \dots, u_n) = f(A^{-1}B) \in L\{u_1, \dots, u_n\}$. It has the property that $F(\sigma(y_1), \dots, \sigma(y_n)) = 0$, for all $\sigma \in H$ but not all $\sigma \in H'$. Assume we are taking F among all polynomials with the preceding property having the smallest number of nonzero monomials. We can assume that some coefficient of F is 1. For $\tau \in H$, let τF be the polynomial obtained by applying τ to the coefficients of F. Then $(\tau F)(\sigma(y_1), \dots, \sigma(y_n)) = \tau(F((\tau^{-1}\sigma(y_1), \dots, \tau^{-1}\sigma(y_n))) = 0$, for all $\sigma \in H$. So, $F - \tau F$ is shorter than F and vanishes for $(\sigma(y_1), \dots, \sigma(y_n))$ for all $\sigma \in H$. By the minimality assumption, it must vanish for $(\sigma(y_1), \dots, \sigma(y_n))$, for all $\sigma \in H'$. If $F - \tau F$ is not identically zero, we can find an element $a \in L$ such that $F - a(F - \tau F)$ is shorter than F and has the same property as F. So $F - \tau F \equiv 0$, for all $\tau \in H$, which means that the coefficients of F are H-invariant. Therefore, F has coefficients in $L^H = L^{H'}$. Now, for $\sigma \in H'$, $F(\sigma(y_1), \dots, \sigma(y_n)) = (\sigma F)(\sigma(y_1), \dots, \sigma(y_n)) = \sigma(F(y_1, \dots, y_n)) = 0$. This contradiction completes the proof. □

Proposition 6.3.2. *Let $K \subset L$ be a differential field extension with differential Galois group $G = G(L|K)$.*

a) If H is a normal subgroup of G, then L^H is G-stable.

b) If F is an intermediate differential field of the extension, which is G-stable, then $G(L|F)$ is a normal subgroup of G. Moreover the restriction morphism

$$\begin{array}{ccc} G(L|K) & \to & G(F|K) \\ \sigma & \mapsto & \sigma_{|F} \end{array}$$

induces an isomorphism from the quotient $G/G(L|F)$ into the group of all differential K-automorphisms of F which can be extended to L.

Proof. a) For $\sigma \in G$, $a \in L^H$, we want to see that $\sigma a \in L^H$. If $\tau \in H$, we have $\tau\sigma a = \sigma a \Leftrightarrow \sigma^{-1}\tau\sigma a = a$ and this last equality is true as $a \in L^H$ and $\sigma^{-1}\tau\sigma \in H$, by the normality of H.

b) To see that $G(L|F)$ is normal in G, we must see that for $\sigma \in G$, $\tau \in G(L|F)$, $\sigma^{-1}\tau\sigma$ belongs to $G(L|F)$, i.e. it fixes every element $a \in F$. Now $\sigma^{-1}\tau\sigma a = a \Leftrightarrow \tau\sigma a = \sigma a$ and this last equality is true since $\sigma a \in F$ because F is G-stable. Now as F is G-stable, we can define a morphism $\varphi : G(L|K) \to G(F|K)$ by $\sigma \mapsto \sigma_{|F}$. The kernel of φ is $G(L|F)$ and its image consists of those differential K-automorphisms of F which can be extended to L. □

Definition 6.3.3. We shall call an extension of differential fields $K \subset L$ *normal* if for each $x \in L \setminus K$, there exists an element $\sigma \in G(L|K)$ such that $\sigma(x) \neq x$.

Proposition 6.3.4. *Let $K \subset L$ be a Picard-Vessiot extension, $G := G(L|K)$ its differential Galois group.*

a) Let H be a closed subgroup of G. If H is normal in G, then the differential field extension $K \subset F := L^H$ is normal.

b) Let F be a differential field with $K \subset F \subset L$. If $K \subset F$ is a Picard-Vessiot extension, then the subgroup $H = G(L|F)$ is normal in $G(L|K)$. In this case, the restriction morphism

$$\begin{array}{ccc} G(L|K) & \to & G(F|K) \\ \sigma & \mapsto & \sigma_{|F} \end{array}$$

induces an isomorphism $G(L|K)/G(L|F) \simeq G(F|K)$.

Proof. a) By Proposition 6.1.2, for $x \in F \setminus K$, there exists $\sigma \in G$ such that $\sigma x \neq x$. By Proposition 6.3.2 a), we know that F is G-stable; hence $\sigma_{|F}$ is an automorphism of F.

b) By Corollary 5.6.7, F is G-stable. Then by Proposition 6.3.2 b), $H = G(L|F)$ is a normal subgroup of $G = G(L|K)$.

For the last part, taking into account Proposition 6.3.2 b), it only remains to prove that the image of the restriction morphism is the whole group $G(F|K)$ which comes from Proposition 6.1.2 b). □

The next proposition establishes the most difficult part of the fundamental theorem, namely that the intermediate field F corresponding to a normal subgroup of G is a Picard-Vessiot extension of K. This result is not proved in Kaplansky's book [**K**], which refers to a paper by Kolchin [**Ko2**]. In fact, Kolchin establishes the fundamental theorem for the larger class of strongly normal extensions and characterizes Picard-Vessiot extensions as strongly normal extensions whose differential Galois group is a linear algebraic group. Our proof is inspired by [**P-S1**] and [**Ż**], but not all details of it can be found there. The proof given in [**M**] uses a different algebra T.

Proposition 6.3.5. *Let $K \subset L$ be a Picard-Vessiot extension, $G(L|K)$ its differential Galois group. If H is a normal closed subgroup of $G(L|K)$, then the extension $K \subset L^H$ is a Picard-Vessiot extension.*

Proof. Let us first explain the idea of the proof. Assume that we have a finitely generated K-subalgebra T of L satisfying the following conditions.

a) T is G-stable and its fraction field $Fr(T)$ is equal to L.

b) For each $t \in T$, the C-vector space generated by $\{\sigma t : \sigma \in G\}$ is finite dimensional.

c) The subalgebra $T^H = \{t \in T : \sigma t = t, \forall \sigma \in H\}$ is a finitely generated K-algebra.

d) $F := L^H$ is the fraction field $Fr(T^H)$ of T^H.

With all these assumptions, let us prove that T^H is generated over K by the space of solutions of a homogeneous linear differential equation with coefficients in K. First let us observe that, as $H \lhd G$, T^H is G-stable, i.e. $\tau(T^H) = T^H$, for all $\tau \in G$. Indeed, let $t \in T^H, \tau \in G$. We want to see that $\tau t \in T^H$. For $\sigma \in H$, we have $\sigma\tau t = \tau t \Leftrightarrow (\tau^{-1}\sigma\tau)t = t$ and the last equality is true as the normality of H implies $\tau^{-1}\sigma\tau \in H$. Thus T^H is a G-stable subalgebra of T and the restriction of the action of G to T^H gives an action of the quotient group G/H on T^H.

We now take a finite-dimensional subspace $V_1 \subset T^H$ over C which generates T^H as a K-algebra and which is G-stable. Note that such a V_1 exists by conditions b) and c). Let $z_1, \dots, z_m$ be a basis of V_1; then the wrońskian $W(z_1, \dots, z_m)$ is not zero. The differential equation in Z

$$\frac{W(Z, z_1, \dots, z_m)}{W(z_1, \dots, z_m)} = 0$$

is satisfied by any $z \in V_1$. Now, by expanding the determinant in the numerator with respect to the first column, we see that each coefficient of the equation is a quotient of two determinants and that all these determinants are multiplied by the same factor $\det \sigma_{|V_1}$ under the action of the element $\sigma \in G$. So these coefficients are fixed by the action of G and so, by using Corollary 6.1.3, we see that they belong to K. Thus $T^H = K\langle V_1 \rangle$, where V_1 is the space of solutions of a linear differential equation with coefficients in K. Therefore $F = L^H = Fr(T^H)$ is a Picard-Vessiot extension of K.

We can assume that L is the Picard-Vessiot extension constructed in Theorem 5.6.5. Let T be the K-algebra R/P considered in the construction. We shall prove that T satisfies the conditions stated above.

a) By construction G acts on T and the fraction field $Fr(T)$ of T is equal to L.

b) Taking into account Remark 6.2.4, we can apply Lemma 3.4.2a) and obtain that the orbit of an element $t \in T$ by the action of G generates a finite dimensional C-vector space.

c) We consider the isomorphism of G-modules given by Proposition 6.2.5 and restrict the action to the subgroup H. The group H acts on both $\overline{K} \otimes_K T$ and $\overline{K} \otimes_C C[G]$ by acting on the second factor. We then have $\overline{K} \otimes_K T^H \simeq \overline{K} \otimes_C C[G]^H$. By Proposition 3.7.8, $C[G]^H \simeq C[G/H]$ as C-algebras. Now $C[G/H]$ is a finitely generated C-algebra and so $\overline{K} \otimes_K T^H$ is a finitely generated $\overline{K}$-algebra. Now we apply the following two lemmas to obtain that T^H is a finitely generated K-algebra.

Lemma 6.3.6. *Let K be a field, $\overline{K}$ an algebraic closure of K and A a K-algebra. If $\overline{K} \otimes_K A$ is a finitely generated $\overline{K}$-algebra, then there exists a finite extension $\widetilde{K}$ of K such that $\widetilde{K} \otimes_K A$ is a finitely generated $\widetilde{K}$-algebra.*

Proof. Let $\{v_s\}_{s \in S}$ be a K-basis of $\overline{K}$ and let $\{\lambda_i \otimes a_i\}_{i=1,\dots,n}$ generate $\overline{K} \otimes_K A$ as a $\overline{K}$-algebra. If we write down the elements λ_i in the K-basis of $\overline{K}$, only the v_s's with s in some finite subset S' of S are involved. We take $\widetilde{K} = K(\{v_s\}_{s \in S'})$. Then the elements $\{v_s \otimes a_i\}_{s \in S', i=1,\dots,n}$ generate $\widetilde{K} \otimes_K A$ as a $\widetilde{K}$-algebra. □

Lemma 6.3.7. *Let K be a field, A a finitely generated K-algebra, and let U be a finite group of automorphisms of A. Then the subalgebra A^U of A fixed by the action of U, i.e. $A^U = \{a \in A : \sigma a = a, \forall \sigma \in U\}$, is a finitely generated K-algebra.*

Proof. For each element $a \in A$, let us define

$$S(a) = \frac{1}{N} \sum_{\sigma \in U} \sigma a,$$

where $N = |U|$, and let us consider the polynomial

$$P_a(T) = \prod_{\sigma \in U} (T - \sigma a) = T^N + \sum_{i=1}^{N} (-1)^i a_i T^{N-i}.$$

The coefficients a_i are the symmetric functions in the roots of $P_a(T)$ and by the Newton formulae can be expressed in terms of the $S(a^i)$, $i = 1, \dots, N$. Let $u_1, \dots, u_m$ now generate A as a K-algebra. We consider the subalgebra B of A^U generated by the elements $S(u_i^j), i = 1, \dots, m, j = 1, \dots, N$. We have $P_{u_i}(u_i) = 0$ and so u_i^N can be written as a lincar combination of $1, \dots, u_i^{N-1}$ with coefficients in B. Hence each monomial $u_1^{a_1} \dots u_m^{a_m}$ can be written in terms of monomials $u_1^{a_1} \dots u_m^{a_m}$, with $a_i < N$ and coefficients in B. Therefore each element $a \in A$ can be written in the form

$$a = \sum_{a_i < N} \varphi_{a_1 \dots a_m} u_1^{a_1} \dots u_m^{a_m},$$

with $\varphi_{a_1 \dots a_m} \in B$. Now, if $a \in A^U$, we have

$$a = S(a) = \sum_{a_i < N} \varphi_{a_1 \dots a_m} S(u_1^{a_1} \dots u_m^{a_m}).$$

Thus A^U can be generated over K by the finite set

$$\{S(u_1^{a_1} \dots u_m^{a_m})\}_{a_i < N} \cup \{S(u_i^N)\}_{i=1,\dots,m}.$$

□

Now by applying Lemma 6.3.6 to $\overline{K} \otimes_K T^H$, we obtain that $\widetilde{K} \otimes_K T^H$ is a finitely generated $\widetilde{K}$-algebra for some finite extension $K \subset \widetilde{K}$ and then also a finitely generated K-algebra. Now we can assume that the extension $K \subset \widetilde{K}$ is normal and consider the Galois group $U = \mathrm{Gal}(\widetilde{K}|K)$ acting on $\widetilde{K} \otimes_K A$ on the left factor. By applying Lemma 6.3.7, we can conclude that $T^H \simeq K \otimes_K T^H \simeq \widetilde{K}^U \otimes_K T^H \simeq (\widetilde{K} \otimes_K T^H)^U$ is a finitely generated K-algebra.

d) We now prove that L^H is the fraction field of T^H.

Let $a \in L^H \setminus \{0\}$. We want to write a as a quotient of elements in T^H. We consider the ideal $J = \{t \in T : ta \in T\}$ of denominators of a. Since a is H-invariant, J is H-stable, i.e. $HJ = J$. Let $s \in J \setminus \{0\}$. Taking into account Remark 6.2.4, we can apply Lemma 3.4.2a) and obtain that the elements $\tau s, \tau \in H$ generate a finite dimensional vector space E over C. Let $s_1, \dots, s_p$ be a basis of E and $w = W(s_1, \dots, s_p)$ be the wrońskian. By expanding the determinant with respect to the first row, we see that $w \in J$. We have $\tau w = \det(\tau_{|E}) \cdot w$, for all $\tau \in H$. We note that $\tau \mapsto \det(\tau_{|E})$ defines a character χ of H; hence w is a semi-invariant with weight χ. (See Section 3.6.) Let $t = wa$. It belongs to T, because $w \in J$, and is a semi-invariant with the same weight as w, because a is H-invariant. So a can be written as t/w the quotient of two semi-invariants. If we find a semi-invariant u with weight $1/\chi$, then we would have $a = (tu)/(wu)$ the quotient of two invariants as desired. We consider the subalgebra of T consisting of the semi-invariants of weight $1/\chi$, that is, $T_{1/\chi} = \{t \in T : \tau t = t/\chi(\tau)\,, \forall \tau \in H\}$. We want to prove $T_{1/\chi} \neq 0$.

To this end, we first consider the action of H on the coordinate ring $C[G]$ of the algebraic group G and prove $C[G]_\eta \neq 0$, for each character η of H. Let us denote $X(H)$ the character group of the group H. Let H_0 be the intersection of the kernels of all characters of H. It is a normal subgroup of H and contains the commutator subgroup of H, so H/H_0 is commutative. By Theorem 4.3.6, H/H_0 is isomorphic to the direct product of its closed subgroups $(H/H_0)_s = \{h \in H/H_0 : h \text{ is semisimple}\}$ and $(H/H_0)_u = \{h \in H/H_0 : h \text{ is unipotent}\}$. By Lemma 4.3.5, $(H/H_0)_u$ is conjugate to a subgroup of the upper triangular unipotent group $\mathrm{U}(n, C)$. Hence by Exercise 10 in chapter 3, $(H/H_0)_u$ does not have nontrivial characters. We then have $X(H) = X(H/H_0) = X((H/H_0)_s)$. We write H' for $(H/H_0)_s$. If η is a character of H', we have $\eta \in C[H']$ and moreover, for each $x, y \in H'$, we have $(x.\eta)(y) = \eta(xy) = \eta(x)\eta(y)$ which gives $x.\eta = \eta(x)\eta$, so η is a semi-invariant of weight η and we get $C[H']_\eta \neq 0$. Now the inclusion $H' \hookrightarrow G/H_0$ corresponds to an epimorphism between the coordinate rings $\pi : C[G/H_0] \to C[H']$. We want to see that $\pi_{|C[G/H_0]_\eta} : C[G/H_0]_\eta \to C[H']_\eta$ is also an epimorphism. Let a be a nonzero element in $C[H']_\eta$. Let $\alpha \in C[G/H_0]$ such that $\pi(\alpha) = a$. By Lemma 3.4.2a), there exists a finite dimensional H'-stable subspace E_1 of $C[G/H_0]$ containing α. As H' is semisimple and commutative, it is diagonalizable, i.e. conjugate in the general linear group to a subgroup of the group of diagonal matrices (cf. Lemma 4.3.5). Therefore the representation of H' on E_1 diagonalizes in a certain basis $\alpha_1, \cdots, \alpha_p$. We can choose it such that $\alpha_1, \cdots, \alpha_l$, with $l < p$ are a basis of $E_1 \cap \operatorname{Ker} \pi$. We have $\alpha = \sum_{j=1}^p c_j \alpha_j \Rightarrow \tau(\alpha) = \sum_{j=1}^p c_j \eta_j(\tau) \alpha_j$, then $\pi(\tau(\alpha)) = \sum_{j=1}^p c_j \eta_j(\tau) \pi(\alpha_j)$ and, on the other hand, $\pi(\tau(\alpha)) = \tau(\pi(\alpha)) = \tau(a) = \eta(\tau) \sum_{j=1}^p c_j \pi(\alpha_j)$. We have $c_j \neq 0$ for some $j > l$ and so $\eta_j(\tau) = \eta(\tau)$ which gives that α_j is a semi-invariant with weight η. We then obtain $0 \neq C[G/H_0]_\eta \subset C[G]_\eta$.

Now we again consider the isomorphism of G-modules given by Proposition 6.2.5 with action restricted to the subgroup H. As the group H acts on both $\overline{K} \otimes_K T$ and $\overline{K} \otimes_C C[G]$ by acting on the second factor, we have $C[G]_{1/\chi} \neq 0 \Rightarrow (\overline{K} \otimes_C C[G])_{1/\chi} \neq 0 \Rightarrow (\overline{K} \otimes_K T)_{1/\chi} \neq 0 \Rightarrow T_{1/\chi} \neq 0$. To obtain the last implication, we use the fact that if $t \in \overline{K} \otimes_K T$, we have $t \in \widetilde{K} \otimes_K T$, for some finite extension $\widetilde{K}$ of K. We can assume that $K \subset \widetilde{K}$ is a normal extension and take $U = G(\widetilde{K}|K)$. Then, if $t \in (\widetilde{K} \otimes_K T)_{1/\chi}$, the element $\sum_{\sigma \in U} \sigma t$ is a semi-invariant with weight $1/\chi$ (as H acts in $\widetilde{K} \otimes_K T$ by acting on the right factor and U by acting on the left factor, both actions commute) and belongs to $K \otimes_K T \simeq T$.

□

Now Propositions 6.3.1, 6.3.4, and 6.3.5 together establish the fundamental theorem of Picard-Vessiot theory.

Theorem 6.3.8 (Fundamental Theorem). *Let $K \subset L$ be a Picard-Vessiot extension, $G(L|K)$ its differential Galois group.*

(1) The correspondences

$$H \mapsto L^H \quad , \quad F \mapsto G(L|F)$$

define inclusion inverting mutually inverse bijective maps between the set of Zariski closed subgroups H of $G(L|K)$ and the set of differential fields F with $K \subset F \subset L$.

(2) The intermediate differential field F is a Picard-Vessiot extension of K if and only if the subgroup $H = G(L|F)$ is normal in $G(L|K)$. In this case, the restriction morphism

$$\begin{array}{ccc} G(L|K) & \to & G(F|K) \\ \sigma & \mapsto & \sigma_{|F} \end{array}$$

induces an isomorphism $G(L|K)/G(L|F) \simeq G(F|K)$.

6.4. Liouville extensions

The aim of this chapter is to characterize linear differential equations solvable by quadratures. This is the analogue of the characterization of algebraic equations solvable by radicals.

Definition 6.4.1. A differential field extension $K \subset L$ is called a *Liouville extension* if there exists a chain of intermediate differential fields $K = F_1 \subset F_2 \subset \cdots \subset F_n = L$ such that F_{i+1} is obtained from F_i either by adjunction of an integral or by adjunction of the exponential of an integral.

Proposition 6.4.2. *Let L be a Liouville extension of the differential field K, having the same field of constants as K. Then the differential Galois group $G(L|K)$ of L over K is solvable.*

Proof. We assume that the extension $K \subset L$ has a chain of intermediate differential fields as in Definition 6.4.1. From Examples 6.1.4 and 6.1.5, we obtain that $K \subset F_2$ is a Picard-Vessiot extension with commutative differential Galois group. By Corollary 5.6.7, every K-differential automorphism of L sends F_2 onto itself. By Proposition 6.3.2 b), $G(L|F_2)$ is a normal subgroup of $G(L|K)$ and $G(L|K)/G(L|F_2)$ is a subgroup of $G(F_2|K)$, hence commutative. By iteration, we obtain that $G(L|K)$ is solvable. □

The next proposition is the first step for a converse of Proposition 6.4.2. In fact we shall consider generalized Liouville extensions, also admitting algebraic extensions as constructing blocks.

Proposition 6.4.3. *Let $K \subset L$ be a normal extension of differential fields. Assume that there exist elements $u_1, \dots, u_n \in L$ such that for every differential automorphism σ of L we have*

$$\sigma\, u_j = a_{1j}\, u_1 + \cdots + a_{j-1,j}\, u_{j-1} + a_{jj}\, u_j\,,\ j = 1, \dots, n, \tag{6.2}$$

with a_{ij} constants in L (depending on σ). Then $K\langle u_1, \dots, u_n\rangle$ is a Liouville extension of K.

Proof. The first of the equations (6.2) is $\sigma u_1 = a_{11} u_1$. Differentiating, we obtain $\sigma u_1' = a_{11} u_1'$ and so u_1'/u_1 is invariant under each σ. (We can assume $u_1 \neq 0$ for otherwise it could simply be suppressed.) By the normality of $K \subset L$, we obtain $u_1'/u_1 \in K$. Hence the adjunction of u_1 to K is the adjunction of an exponential. Next we divide each of the next $n-1$ equations by the equation $\sigma u_1 = a_{11} u_1$ and differentiate. The result is

$$\sigma \left(\frac{u_j}{u_1}\right)' = \frac{a_{2j}}{a_{11}} \left(\frac{u_2}{u_1}\right)' + \cdots + \frac{a_{j-1,j}}{a_{11}} \left(\frac{u_{j-1}}{u_1}\right)' + \frac{a_{jj}}{a_{11}} \left(\frac{u_j}{u_1}\right)'.$$

This is a set of equations of the same form as (6.2) in the elements $(u_j/u_1)'$, with $j = 2, \dots, n$. By induction on n, the adjunction of $(u_j/u_1)'$ to K yields a Liouville extension. Then adjoining u_j/u_1 themselves means adjoining integrals. □

6.5. Generalized Liouville extensions

Definition 6.5.1. A differential field extension $K \subset L$ is called a *generalized Liouville extension* if there exists a chain of intermediate differential fields $K = F_1 \subset F_2 \subset \cdots \subset F_n = L$ such that either F_{i+1} is obtained from F_i by adjunction of an integral or by adjunction of the exponential of an integral or F_{i+1} is algebraic over F_i.

A solution of a differential equation defined over the differential field K is called *Liouvillian* if it is contained in a generalized Liouville extension of K.

Theorem 6.5.2. *Let K be a differential field with algebraically closed field of constants C. Let L be a Picard-Vessiot extension of K. Assume that the identity component G_0 of $G = G(L|K)$ is solvable. Then L can be obtained from K by a finite normal extension followed by a Liouville extension.*

Proof. Let $F = L^{G_0}$. We know by Proposition 3.2.1 that G_0 is a normal subgroup of G of finite index. Then $K \subset F$ is a finite normal extension and $G(L|F) \simeq G_0$. Then by Lie-Kolchin Theorem 4.4.6, we can apply Proposition 6.4.3 and obtain that $F \subset L$ is a Liouville extension. □

To prove an inverse to this theorem we shall use the following lemma.

Lemma 6.5.3. *Let K be a differential field with algebraically closed field of constants C. Let L be a Picard-Vessiot extension of K. Let $L_1 = L\langle z\rangle$ be an extension of L with no new constants. Write $K_1 = K\langle z\rangle$. Then $K_1 \subset L_1$ is a Picard-Vessiot extension and its differential Galois group is isomorphic to $G(L|L \cap K_1)$.*

Proof. It is clear that $K_1 \subset L_1$ is a Picard-Vessiot extension as both fields have the same field of constants and the extension is generated by the solutions of the differential equation associated to the Picard-Vessiot extension $K \subset L$. By Corollary 5.6.7, any K-differential automorphism of L_1 sends L onto itself. Thus restriction to L gives a morphism $\varphi : G(L_1|K_1) \rightarrow G(L|K)$. An automorphism of L_1 in $\operatorname{Ker}\varphi$ fixes both K_1 and L and so is the identity. Hence φ is injective and $G(L_1|K_1)$ is isomorphic to a closed subgroup of $G(L|K)$. The corresponding intermediate field of the extension $K \subset L$ is $L \cap K_1$ and by the Fundamental Theorem 6.3.8 we get $G(L_1|K_1) \simeq G(L|L \cap K_1)$. □

Theorem 6.5.4. *Let K be a differential field with algebraically closed field of constants C. Let L be a Picard-Vessiot extension of K. Assume that L can be embedded in a differential field M which is a generalized Liouville* *extension of K with no new constants. Then the identity component G_0 of $G = G(L|K)$ is solvable (whence by Theorem 6.5.2, L can be obtained from K by a finite normal extension, followed by a Liouville extension).*

Proof. We make an induction on the number of steps in the chain from K to M. Let $K\langle z\rangle$ be the first step. Then, by induction, the differential Galois group of $L\langle z\rangle$ over $K\langle z\rangle$ has a solvable component of the identity. By Lemma 6.5.3, this group is isomorphic to the subgroup H of G corresponding to $L \cap K\langle z\rangle$. Assume that z is algebraic over K. Then, H has finite index in G. In this case, by Proposition 3.2.1 b), $G^0 = H^0$, hence solvable. If z is either an integral or the exponential of an integral, by Examples 6.1.4 and 6.1.5, $K\langle z\rangle$ is a Picard-Vessiot extension of K with commutative Galois group. Thus all differential fields between K and $K\langle z\rangle$ are normal over K. In particular, $L \cap K\langle z\rangle$ is normal over K with a commutative differential Galois group. Thus H is normal in G with G/H commutative. So by Lemma 4.4.4, the identity component G^0 of G is solvable. □

Exercises

(1) Given the differential equation $\mathcal{L}(Y) = Y^{(n)} + a_{n-1}Y^{(n-1)} + \cdots + a_1Y' + a_0Y = 0$, prove that the change $Y = vZ$, where v is a solution of $nv' + a_{n-1} = 0$, gives a differential equation $\widetilde{\mathcal{L}}(Z) = 0$ without term in $Z^{(n-1)}$. Prove that all solutions to $\mathcal{L}(Y) = 0$ are Liouvillian if and only if all solutions to $\widetilde{\mathcal{L}}(Z) = 0$ are Liouvillian.

(2) Let $\mathcal{L}(Y) := Y^{(n)} + a_{n-1}Y^{(n-1)} + \cdots + a_1Y' + a_0Y$ and let W denote the wrońskian determinant of a fundamental set of solutions of $\mathcal{L}(Y) = 0$. Prove that if $a_{n-1} = 0$, then W is constant and $\text{Gal}(\mathcal{L})$ is contained in the special linear group.

(3) Give an example of a finite Galois extension of differential fields (with nonalgebraically closed field of constants) which is not Picard-Vessiot.

(4) Prove the claims in Remark 6.2.3.

(5) Determine the differential Galois group of the extension $\mathbb{C} \subset \mathbb{C}(t)$, where the derivation in $\mathbb{C}(t)$ is given by $t' = 1$.

(6) Determine the differential Galois group of a homogeneous linear differential equation with coefficients in $\mathbb{C}$
a) over $\mathbb{C}(t)$ (with derivation given by $t' = 1$),
b) over $\mathbb{C}$.

(7) Let K_0 be a differential field with field of constants C. Let $L := K_0\langle y_1, \ldots, y_n \rangle$ be the field of differential rational functions in the differential variables $y_1, \ldots, y_n$ over K_0. The action of a matrix $\sigma \in \text{GL}(n, C)$ on the vector $(y_1, \ldots, y_n)$ extends to an action on L by differential K_0-automorphisms. Let $K := L^{\text{GL}(n,C)}$ be the field fixed by this action. Prove that $K \subset L$ is a Picard-Vessiot extension with differential Galois group $\text{GL}(n, C)$.

(8) Let $\mathcal{L} \in K[D]$ be a monic differential operator of degree n and consider the differential equation $\mathcal{L}(Y) = 0$. Let L be a Picard-Vessiot extension for $\mathcal{L}(Y) = 0$ and $G = G(L|K)$ its differential Galois group. If V is the C_K-vector space of solutions to $\mathcal{L}(Y) = 0$ in L, then G acts on V, giving a faithful representation of order n of G over C_K. Prove that this representation has an invariant subspace of dimension m if and only if $\mathcal{L}$ has a right factor in $K[D]$ of degree m.
Hint: If U is an invariant subspace of dimension m, $y_1, \ldots, y_m$ a C_K-basis of U, prove that

$$\mathcal{L}_1(Y) := \frac{W(Y, y_1, \ldots, y_m)}{W(y_1, \ldots, y_m)} = 0$$

is a differential equation with coefficients in K and use right division of differential operators.

(9) A *Bernouilli equation* is an ordinary differential equation of the form

$$Y' + pY = qY^n \tag{6.3}$$

with $n \neq 0, 1$. Show that the change $Z = Y^{-(n-1)}$ transforms (6.3) into a first order linear differential equation and give the expression of the general solution of the Bernoulli equation.

(10) A *Riccati equation* is an ordinary differential equation of the form

$$Y' = q_0 + q_1 Y + q_2 Y^2. \tag{6.4}$$

a) Prove that, if $q_2 \neq 0$, the change $Z = q_2 Y$ gives a Riccati equation of the form

$$Z' = p_0 + p_1 Z + Z^2.$$

b) Prove that if u is a solution of the second order linear differential equation

$$U'' - p_1 U' + p_0 U = 0,$$

then $y = -u'/q_2 u$ is a solution of (6.4).

c) Prove that, if a particular solution y_1 of (6.4) is known, the substitution $Y = y_1 + U$ leads to a Bernouilli equation in U and so we can obtain the general solution of the Riccati equation.

(11) Taking into account Exercise 20 in chapter 5, prove that if a homogeneous linear differential equation of order 2 has a Liouvillian solution, then all its solutions are Liouvillian.

Chapter 7

Differential Equations over $\mathbb{C}(z)$

In this chapter we consider linear differential equations over the field $\mathbb{C}(z)$ of rational functions in the variable z over the field $\mathbb{C}$ of complex numbers. We give some classical results related to local solutions of differential equations defined over $\mathbb{C}(z)$. In the last part, we present Kovacic's algorithm which determines explicitly the Liouvillian solutions to differential equations of order 2.

We assume that the reader is familiar with the basic concepts of complex analysis in one variable, as presented in [**Al**].

7.1. Fuchsian differential equations

For a linear differential equation

$$Y^{(n)} + a_1(z)\,Y^{(n-1)} + \cdots + a_{n-1}(z)\,Y' + a_n(z)\,Y = 0, \tag{7.1}$$

with $a_i(z) \in \mathbb{C}(z)$, a point P in $\mathbb{C}$ is called *regular* if the functions a_i have no pole in P; otherwise P is called *singular*. To consider whether ∞ is a regular or singular point, we make the substitution $z = x^{-1}$ in (7.1). If $x = 0$ is a regular (resp. singular) point for the new equation, then $z = \infty$ is a regular (resp. singular) point for (7.1).

Definition 7.1.1. If $P \in \mathbb{C}$ (resp. $P = \infty$) is a singular point for (7.1), we consider the limit $\lim_{z\to P}(z-P)^i a_i(z)$ (resp. $\lim_{z\to\infty} z^i a_i(z)$). If this limit exists and is finite for $i = 1, \dots, n$, the point P is a *regular singular point* for (7.1).

The equation (7.1) is called *Fuchsian* if all points in $\mathbb{P}^1(\mathbb{C})$ are regular or regular singular.

Proposition 7.1.2. *For a Fuchsian differential equation with regular singular points $P_1, \ldots, P_\nu$ in $\mathbb{C}$, the coefficients $a_i(z)$ are of the form*

$$a_i(z) = \frac{A_i(z)}{\prod_{s=1}^{\nu}(z-P_s)^i},$$

where $A_i(z)$ is a polynomial of degree $\leq i(\nu-1)$. In particular

$$a_1(z) = \sum_{s=1}^{\nu} \frac{A_{1s}}{z-P_s},$$

where A_{1s} are constants.

Proof. By the definition of regular singular points, we have

$$a_i(z) = \prod_{s=1}^{\nu}(z-P_s)^{-i} A_i(z),$$

where $A_i(z)$ is an entire function of z, as there are no other singular points in $\mathbb{C}$. We now consider the behavior of these coefficients at infinity. The point at infinity is at most a pole of $a_i(z)$; consequently, $A_i(z)$ is a polynomial in z and $\lim_{z\to\infty} z^i a_i(z)$ finite implies that its degree is $\leq i(\nu-1)$. In particular

$$a_1(z) = \prod_{s=1}^{\nu}(z-P_s)^{-1}A_1(z) = \sum_{s=1}^{\nu}((z-P_s)^{-1}A_{1s}),$$

with $A_{1s} = A_1(P_s)/\prod_{t\neq s}(P_s-P_t)$. □

Let us now assume that 0 is a regular singular point for Equation (7.1). We write the equation in terms of the differential operator $D = z\dfrac{d}{dz}$. We have

$$z^r \frac{d^r}{dz^r} = D(D-1)\cdots(D-r+1) \tag{7.2}$$

and obtain a differential equation of the form

$$F(D,z)(Y) := D^nY + b_1(z)D^{n-1}Y + \cdots + b_{n-1}(z)DY + b_n(z)Y = 0, \tag{7.3}$$

where the functions $b_i(z)$ are holomorphic in a neighborhood of $z=0$. (See Exercise 18 in chapter 5.)

We shall now see that we can obtain formally the solutions of the differential equation (7.1) as power series in the neighborhood of a regular point. By a change of variable, we can assume that this point is 0.

Theorem 7.1.3 (Cauchy theorem). *Let 0 be a regular point of (7.1). Then there exist n power series in z, $f_1, \dots, f_n$, which are solutions of (7.1), linearly independent over $\mathbb{C}$, with positive convergence radius. Moreover, every power series which is a solution of (7.1) is a linear combination of $f_1, \dots, f_n$ with coefficients in $\mathbb{C}$.*

Proof. We look for a solution in power series $y = \sum_{k\geq 0} c_k z^k$. Multiplying (7.1) by z^n and using (7.2), we obtain

$$D(D-1)\cdots(D-n+1)Y + \sum_{i=1}^{n} z^i a_i(z) D(D-1)\cdots(D-(n-i)+1)Y = 0.$$

We write $z^i a_i(z) = \sum_{j=i}^{\infty} a_{ij} z^j$ and, for each $j \geq 1$, we set

$$Q_j(X) := \sum_{i=1}^{j} a_{ij} X(X-1)\cdots(X-(n-i)+1).$$

By substituting y in the equation, we obtain the recurrence relation

$$k(k-1)\cdots(k-n+1)c_k + \sum_{j=1}^{k} Q_j(k) c_{k-j} = 0.$$

As $Q_j(k-j) = 0$, for $1 \leq j \leq k < n$, the recurrence is trivial for $k \leq n-1$. So we can fix $c_0, \cdots, c_{n-1}$ arbitrarily and the coefficients c_k with $k \geq n$ are determined by the recurrence relation. We then obtain n linearly independent solutions of (7.1) which are a basis of the vector space of solutions.

It remains to be seen that every solution in power series has a positive radius of convergence. To this end, we choose $C > 1$, satisfying the following conditions.

(1) $|Q_j(k)| < C^j k^{n-1}$ for all j, k,

(2) $|c_j| < C^{2j+1}$ for $j = 0, \cdots n-1$,

(3) $(k(k-1)\cdots(k-n+1))^{-1} < C/k^n$ for all $k \geq n$.

We prove by induction over k that $|c_k| < C^{2k+1}$. For $k < n$, it is true by the choice of C. From the recurrence relation, we obtain the inequality

$$|c_k| \leq \frac{C}{k^n} \sum_{j=1}^{k} C^j k^{n-1} |c_{k-j}|.$$

Using the induction hypothesis,

$$|c_k| \le \frac{C}{k^n} k^{n-1} \sum_{j=1}^{k} C^j \cdot C^{2(k-j)+1} < C^{2k+1}.$$

Hence the coefficients c_k are exponentially bounded and the power series has a positive radius of convergence. □

We shall now see how to obtain formally the solutions in the neighborhood of a regular singular point. Let us assume that 0 is a regular singular point of the equation (7.1). We look for solutions of the form

$$y = z^\rho \sum_{k \ge 0} c_k z^k. \tag{7.4}$$

We develop the coefficients of (7.3) in Taylor series, $b_i(z) = \sum_{j=0}^{\infty} b_{ij} z^j$ and set

$$\begin{cases} F_0(D) &= D^n + b_{10} D^{n-1} + b_{20} D^{n-2} + \cdots + b_{n0} \\ F_j(D) &= b_{1j} D^{n-1} + b_{2j} D^{n-2} + \cdots + b_{nj} \quad \text{for} \quad j > 0. \end{cases}$$

The equation can then be written as

$$F(D, z)(Y) = \sum_{j=0}^{\infty} z^j F_j(D)(Y) = 0$$

and substituting y, we obtain

$$\begin{aligned} F(D, z)(y) &= \textstyle\sum_{i=0}^{\infty} \sum_{j=0}^{\infty} z^j F_j(D)(c_i z^{\rho+i}) \\ &= \textstyle\sum_{i=0}^{\infty} \sum_{j=0}^{\infty} z^{\rho+i+j} F_j(\rho+i) c_i \\ &= \textstyle\sum_{k=0}^{\infty} z^{\rho+k} \left[\sum_{i=0}^{k} F_{k-i}(\rho+i) c_i \right] = 0. \end{aligned}$$

This expression vanishes identically if the coefficients c_i satisfy the relations

$$\sum_{i=0}^{k} F_{k-i}(\rho+i) c_i = 0 \quad (k \ge 0). \tag{7.5}$$

In particular, to obtain $c_0 \ne 0$, ρ must be a root of the polynomial equation

$$F_0(X) = X^n + b_{10}X^{n-1} + \cdots + b_{n0} = X^n + b_1(0)X^{n-1} + \cdots + b_n(0) = 0.$$

This equation is called *indicial equation*. Its roots are called *local exponents* at the singular point $z = 0$.

From (7.2), we obtain that the indicial equation at a regular or regular singular point $P \in \mathbb{C}$ in terms of the coefficients of (7.1) is

$$\begin{aligned}&X(X-1)\cdots(X-n+1)\\ &\quad+\textstyle\sum_{k=1}^{n} lim_{z\to P}(z-P)^k a_k(z)X\cdots(X-n+k+1)\\ &\qquad\qquad + lim_{z\to P}(z-P)^n a_n(z) = 0.\end{aligned} \tag{7.6}$$

If ∞ is a regular or regular singular point, the indicial equation at ∞ is

$$\begin{aligned}&X(X+1)\cdots(X+n-1)\\ &\quad+\textstyle\sum_{k=1}^{n}(-1)^k lim_{z\to\infty} z^k a_k(z)X\cdots(X+n-k-1)\\ &\qquad\qquad + (-1)^n lim_{z\to\infty} z^n a_n(z) = 0.\end{aligned} \tag{7.7}$$

Note that at a regular point the local exponents are $0, 1, \cdots, n-1$. The local exponents for a Fuchsian differential equation satisfy the Fuchs relation given in the next proposition.

Proposition 7.1.4 (Fuchs' relation). *If $\rho_1(P), \rho_2(P), \cdots, \rho_n(P)$ are the local exponents at a point $P \in \mathbb{P}^1$ for a Fuchsian differential equation, we have*

$$\sum_{P\in\mathbb{P}^1} (\rho_1(P) + \rho_2(P) + \cdots + \rho_n(P) - \binom{n}{2}) = -2\binom{n}{2},$$

where only the singular points give nonzero summands.

Proof. Let us assume that $P_s, 1 \leq s \leq \nu$, are the singularities of the given equation in $\mathbb{C}$. Using the form for the coefficients of a Fuchsian equation given by Proposition 7.1.2, we compute $\lim_{z\to P_s}(z - P_s)a_1(z) = A_{1s}$ and $\lim_{z\to\infty} z a_1(z) = \sum_{s=1}^{\nu} A_{1s}$. Now by formulae (7.6), (7.7), we obtain

$$\rho_1(P_s) + \rho_2(P_s) + \cdots + \rho_n(P_s) = \binom{n}{2} - A_{1s}$$

and

$$\rho_1(\infty) + \rho_2(\infty) + \cdots + \rho_n(\infty) = -\binom{n}{2} + \sum_{s=1}^{\nu} A_{1s}.$$

Summing up for $P_1, \dots, P_s$ and ∞, we obtain that the sum of all local exponents is equal to $(\nu - 1)\binom{n}{2}$, and this equality can be written as in the statement. $\square$

Now if two independent solutions correspond to the same exponent ρ, by subtracting one from the other, we obtain another solution corresponding to a bigger exponent ρ' which is also a root of the indicial equation. Hence each local exponent gives rise to at most one solution in power series of the form (7.4).

If $F_0(\rho) = 0$, $F_0(\rho + k) \neq 0$ for every positive integer k, we can choose $c_0 \neq 0$ and each of the next coefficients c_k is uniquely determined by the relations (7.5). But if both ρ and $\rho + k$ are local exponents, with k a positive integer, the relation $\sum_{i=0}^{k} F_{k-i}(\rho + i)c_i = 0$ may be incompatible.

In the case in which the indicial equation has multiple roots or roots which differ by an integer number and then we do not have a complete system of solutions of the form (7.4), we can look for solutions in which logarithms appear.

We distribute the local exponents in equivalent classes, obtained by defining two local exponents to be equivalent when they differ by an integer number. Let us now see how to compute the solutions corresponding to one of these classes, formed by h distinct local exponents ρ_i with multiplicities r_i, ordered by ascendent real parts. We look for solutions of the form

$$y = z^{\rho} \sum_{k \geq 0} u_k z^k, \tag{7.8}$$

where the u_k are polynomials in $t := \log z$ of degree smaller than n. Taking into account $D(u_i) = \dfrac{du_i}{dt}$, we obtain

$$\begin{aligned} F(D,z)(y) &= \textstyle\sum_{i=0}^{\infty} \sum_{j=0}^{\infty} z^j F_j(D)(z^{\rho+i} u_i) \\ &= \textstyle\sum_{i=0}^{\infty} \sum_{j=0}^{\infty} z^{\rho+i+j} F_j(D + \rho + i) u_i \\ &= \textstyle\sum_{k=0}^{\infty} z^{\rho+k} \left[\sum_{i=0}^{k} F_{k-i}(D + \rho + i) u_i \right] = 0, \end{aligned}$$

which vanishes identically if the u_i satisfy the relations

$$\sum_{i=0}^{k} F_{k-i}(D+\rho+i)u_i = 0 \quad (k \geq 0).$$

These relations can be seen as a system of linear differential equations in the variable $t = \log z$, of which we are interested in the general solution in polynomials. The first equation can be written as

$$F_0(D+\rho)(u_0) = F_0(\rho)u_0 + \frac{1}{1!}F_0'(\rho)Du_0 + \frac{1}{2!}F_0''(\rho)D^2u_0 + \cdots = 0.$$

This is the *generalized indicial equation.* If u_0 is a polynomial in t, not identically 0, this expression is a polynomial of the same degree unless $F_0(\rho) = 0$. To obtain a solution effectively, ρ must then satisfy the indicial equation. For $\rho = \rho_1$, we have $F_0(D+\rho_1)(u_0) = G_1(D)(D^{r_1}u_0)$ with $G_1(0) \neq 0$, and so u_0 must satisfy the equation $D^{r_1}u_0 = 0$.

Let us assume that we have found the polynomials $u_0, u_1, \cdots, u_{k-1}$. If $F_0(\rho_1 + k) \neq 0$, u_k is uniquely determined as a polynomial whose degree does not exceed the degree of the preceding polynomials by the symbolic formula

$$\begin{aligned} u_k &= -\frac{1}{F_0(D+\rho_1+k)} \textstyle\sum_{i=0}^{k-1} F_{k-i}(D+\rho_1+i)(u_i) \\ &= -(A_0 + A_1D + A_2D^2 + \cdots) \textstyle\sum_{i=0}^{k-1} F_{k-i}(D+\rho_1+i)u_i \qquad (7.9) \\ &= L_k(u_0, u_1, \cdots, u_{k-1}). \end{aligned}$$

But if $k = \rho_i - \rho_1$, we have $F_0(D+\rho_1+k) = F_0(D+\rho_i) = G_i(D)D^{r_i}$, with $G_i(0) \neq 0$ and so, instead of (7.9), we have the relation

$$D^{r_i}u_{\rho_i-\rho_1} = L_k(u_0, u_1, \cdots, u_{k-1}), \quad \text{with} \quad k = \rho_i - \rho_1.$$

The structure of the solution is completely determined by $u_0, \cdots, u_{\rho_h-\rho_1}$ as the remaining u_k are determined by a relation of the form (7.9). We can distinguish the h critical polynomials $U_i := u_{\rho_i-\rho_1}$ and express the remaining ones explicitly in the form

$$u_k = \Lambda_k(U_1, U_2, \cdots, U_h),$$

where the U_i satisfy a system of equations of the form

$$\begin{cases} D^{r_1}U_1 &= 0 \\ f_{21}(D)U_1 + D^{r_2}U_2 &= 0 \\ f_{31}(D)U_1 + f_{32}(D)U_2 + D^{r_3}U_3 &= 0 \\ \dots \end{cases}.$$

We then obtain $\sum_{i=1}^{h} r_i$ linear independent solutions.

We have then seen that we can obtain n independent solutions to the differential equation in the neighborhood of a regular singular point having either the form (7.4) or the form (7.8). It can be proved that they have a positive radius of convergence. (See [**Po**], V.17.)

Remark 7.1.5. A Fuchsian homogeneous linear differential equation of order two with three singular points is completely determined by its singular points and the local exponents at each singular point.

Example 7.1.6. The *Gauss hypergeometric equation*

$$Y'' + \frac{(a+b+1)z - c}{z(z-1)} Y' + \frac{ab}{z(z-1)} Y = 0 \tag{7.10}$$

has three singular points $0, 1, \infty$ which are regular singular. We write down the local exponents in the Riemann scheme:

$$\begin{array}{ccc} 0 & 1 & \infty \\ \hline 0 & 0 & a \\ 1-c & c-a-b & b \end{array}. \tag{7.11}$$

The solution relative to the singular point $z = 0$ and the local exponent 0 is developable in the series

$$\sum_{n=0}^{\infty} \frac{(a)_n (b)_n}{(c)_n n!} z^n,$$

denoted as $F(a, b, c; z)$. (See Exercise 19 in chapter 5.) It may be verified that the series converges for $|z| < 1$, whenever c is not a negative integer and diverges for $|z| > 1$. If a, b, c are real, the series converges for $z = 1$ if $c > a + b$ and diverges if $c \leq a + b$; it converges for $z = -1$ if $c + 1 > a + b$ and diverges if $c + 1 \leq a + b$.

The solution relative to $z = 0$ and the local exponent $1 - c$ is

$$z^{1-c} F(a - c + 1, b - c + 1, 2 - c; z).$$

The two solutions at $z = 1$ are

$$F(a, b, a+b-c+1; 1-z),$$

$$(1-z)^{c-a-b} F(c-a, c-b, c-a-b+1; 1-z).$$

The two solutions at $z = \infty$ are

$$z^{-a} F(a, a-c+1, a-b+1; z^{-1}),$$

$$z^{-b} F(b, b-c+1, b-a+1; z^{-1}).$$

The domain of convergence is $0 < |z| < 2$ for the series in $1-z$ and $z > 1$ for the series in z^{-1}.

Since the equation cannot have more than two linearly independent solutions, there must be dependence relations between the six solutions we have found. It can be proved that

$$F(a, b, c; z) = \frac{\Gamma(c)\Gamma(c-a-b)}{\Gamma(c-a)\Gamma(c-b)} F(a, b, a+b-c+1; 1-z)$$

$$+ \frac{\Gamma(c)\Gamma(a+b-c)}{\Gamma(a)\Gamma(b)} (1-z)^{c-a-b} F(c-a, c-b, c-a-b+1; 1-z),$$

where Γ denotes the Gamma function. (See [**K12**].)

7.2. Monodromy group

Any analytic solution of (7.1) in the neighborhood of a regular point can be analytically continued along any path in $\mathbb{C}$ not passing through any singular point. Let S be the set of singular points of (7.1), $z_0 \in \mathbb{P}^1 \setminus S$. Let $f_1, \ldots, f_n$ be linearly independent analytic solutions in the neighborhood of z_0. Let $\gamma \in \pi_1(\mathbb{P}^1 \setminus S, z_0)$. By analytic continuation along γ, we obtain $\widetilde{f_1}, \ldots, \widetilde{f_n}$ which are solutions of (7.1) as well. We then have a matrix $M(\gamma) \in \mathrm{GL}(n, \mathbb{C})$ such that

$$\begin{pmatrix} \widetilde{f_1} \\ \vdots \\ \widetilde{f_n} \end{pmatrix} = M(\gamma) \begin{pmatrix} f_1 \\ \vdots \\ f_n \end{pmatrix}.$$

The mapping

$$\begin{array}{rccc} \rho: & \pi_1(\mathbb{P}^1 \setminus S) & \to & \mathrm{GL}(n, \mathbb{C}) \\ & \gamma & \mapsto & M(\gamma) \end{array}$$

is a group morphism. Its image M is called *monodromy group of* (7.1). It is determined up to conjugation. Since an element of the differential Galois group of the differential equation (7.1) is determined by the images of a fundamental set of solutions, we can see M as a subgroup of the differential Galois group of the differential equation.

For a Fuchsian differential equation, we have the following relation between the differential Galois group and the monodromy group.

Proposition 7.2.1. *Let us assume that (7.1) is a Fuchsian differential equation. Let G be its differential Galois group, M its monodromy group. Then G is the Zariski closure of M.*

Proof. Let L be a Picard-Vessiot extension for (7.1) over $\mathbb{C}(z)$ and let F denote the subfield of L fixed by M. Then $G(L|F) = \overline{M}$. (See the proof of Proposition 6.3.1.) Thus, by the fundamental theorem of differential Galois theory (see Proposition 6.3.1), it is enough to prove that $F = \mathbb{C}(z)$. Now suppose that $f(z)$ is an element in L fixed by each element of M. It follows that $f(z)$ is a single valued analytic function on $\mathbb{P}^1(\mathbb{C})$ whose singularities belong to the set $\{P_1, \dots, P_s\}$ of singular points of (7.1). Since each P_i is a regular singular point, $f(z)$ approaches a limit as z approaches P_i. Consequently, $f(z)$ has no essential singularities, so it must be meromorphic on $\mathbb{P}^1(\mathbb{C})$ and therefore a rational function of z. □

Remark 7.2.2. We note that Proposition 7.2.1 may fail when (7.1) is not a Fuchsian equation. For example, the monodromy group of the equation $Y' = Y$ is trivial because the solution $y = e^z$ is single valued, but the differential Galois group is $\mathbb{G}_m$. (See Example 6.1.5.) Note that ∞ is a nonregular singularity of the equation.

Example 7.2.3. Let us again consider the *hypergeometric equation*

$$Y'' + \frac{(a+b+1)z - c}{z(z-1)} Y' + \frac{ab}{z(z-1)} Y = 0 \tag{7.12}$$

and its Riemann scheme

$$\begin{array}{ccc} 0 & 1 & \infty \\ \hline 0 & 0 & a \\ 1-c & c-a-b & b \end{array} . \tag{7.13}$$

Let P be a regular point of the equation. With origin in P, we draw three loops in the complex plane g_0, g_1, g_∞, around $0, 1, \infty$, respectively, with $g_0 g_1 g_\infty = 1$. (See Figure 1.) The corresponding monodromy matrices

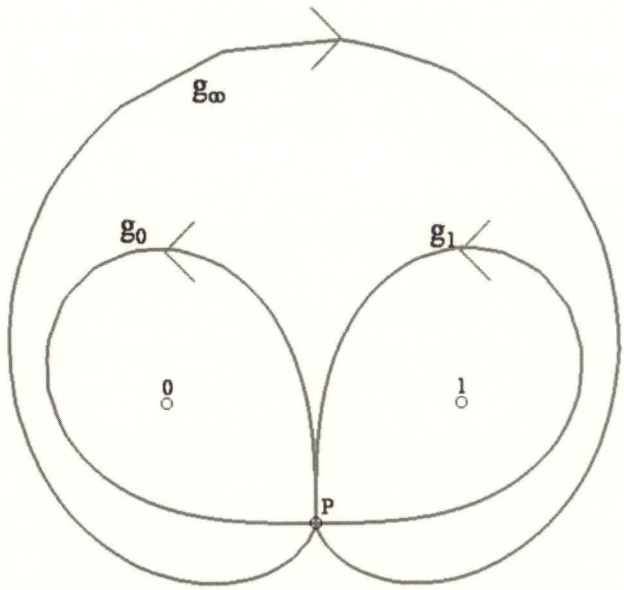

Figure 1. The three loops g_0, g_1, g_∞.

M_0, M_1, M_∞ satisfy $M_0 M_1 M_\infty = 1$ and M_0, M_∞ generate the monodromy group.

From the values of the local exponents given in the Riemann scheme, we have that the eigenvalues of M_0 are $1, e^{2\pi i(1-c)}$, those of M_1, $1, e^{2\pi i(c-a-b)}$, and those of M_∞, $e^{2\pi i a}, e^{2\pi i b}$.

Remark 7.2.4. Any Fuchsian equation of order two with three singular points can be transformed in a hypergeometric equation by means of

- a Möbius transformation $S(z) = \dfrac{az+b}{cz+d}$, $ad - bc \neq 0$ which sends the singular points to $0, 1, \infty$. We then obtain an equation with a Riemann scheme of the form

$$\begin{array}{ccc} 0 & 1 & \infty \\ \hline \alpha & \beta & \gamma \\ \alpha' & \beta' & \gamma' \end{array},$$

where $\alpha + \beta + \gamma + \alpha' + \beta' + \gamma' = 1$ by Fuchs' relation.

- multiplication of the solutions by $z^{-\alpha}(1-z)^{-\beta}$. The obtained Riemann scheme corresponds to a hypergeometric equation with adequate parameters.

7.3. Kovacic's algorithm

The aim of this section is to present Kovacic's algorithm for computing Liouvillian solutions of linear differential equations of order 2 defined over the rational function field in one variable over the field of complex numbers given in [**Kov**]. We give some examples of application of this algorithm which are different than the ones given in [**Kov**]. In particular, the differential equations whose differential Galois group is isomorphic to $2A_4, 2S_4$, or $2A_5$ are taken from [**C-H2**].

By Exercise 1 in chapter 6, we can assume that the equation considered has the form $Y'' = rY$ for some $r \in \mathbb{C}(z)$. Hence the differential Galois group of the equation is an algebraic subgroup of $\mathrm{SL}(2,\mathbb{C})$. (See Exercise 2 in chapter 6.) The starting point of the algorithm is the determination of the closed subgroups of $\mathrm{SL}(2,\mathbb{C})$ given in Theorem 4.6.1. We recall that if a second order homogeneous linear differential equation has a Liouvillian solution, then all its solutions are Liouvillian. (See Exercise 11 in chapter 6.)

It is worth noting that Kovacic's algorithm is essential in the effective application of Morales-Ramis criterion to nonintegrability of Hamiltonian systems. (See **3.** in chapter 8.) It has been applied in [**D-L**] to parametric families of differential equations, in particular to hypergeometric ones, in order to determine when these have Liouvillian solutions.

7.3.1. Determination of the possible cases. We shall consider a differential equation

$$Y'' = rY, \text{ where } r \in \mathbb{C}(z). \tag{7.14}$$

We take $r \notin \mathbb{C}$ to avoid triviality. Regarding the existence of a (nonzero) Liouvillian solution of (7.14), the possible cases are given by the next theorem.

Theorem 7.3.1. *There are precisely four cases that can occur.*

1. *The differential equation (7.14) has a solution of the form $e^{\int \omega}$, where $\omega \in \mathbb{C}(z)$.*
2. *The differential equation (7.14) has a solution of the form $e^{\int \omega}$, where ω is algebraic of degree 2 over $\mathbb{C}(z)$ and case 1 does not hold.*

3. *All solutions of the differential equation (7.14) are algebraic over* $\mathbb{C}(z)$ *and cases 1 and 2 do not hold.*
4. *The differential equation (7.14) has no Liouvillian solution.*

Proof. Let (y_1, y_2) be a fundamental set of solutions of (7.14) and $L = \mathbb{C}(z)\langle y_1, y_2\rangle$. Let $G \subset \mathrm{SL}(2,\mathbb{C})$ be the differential Galois group of (7.14) relative to the basis (y_1, y_2). We shall consider the four cases in Theorem 4.6.1.

1. If G is triangularizable, we may assume that G is triangular. Hence for each $\sigma \in G$, we have $\sigma(y_1) = c_\sigma y_1$ which implies $\sigma(y_1') = c_\sigma y_1'$ and so $\omega = y_1'/y_1 \in L^G = \mathbb{C}(z)$.
2. If G is conjugate to a subgroup of D^+, we may assume that G is a subgroup of D^+. Then, for $\sigma \in G$, we have either $\sigma(y_1) = c_\sigma y_1$, $\sigma(y_2) = c_\sigma^{-1} y_2$ or $\sigma(y_1) = -c_\sigma^{-1} y_2$, $\sigma(y_2) = c_\sigma y_1$. Hence for $\omega = y_1'/y_1$, $\omega_2 = y_2'/y_2$, we have either $\sigma\omega = \omega, \sigma\omega_2 = \omega_2$ or $\sigma\omega = \omega_2, \sigma\omega_2 = \omega$, so ω is quadratic over $\mathbb{C}(z)$.
3. If G is finite, L has only a finite number of $\mathbb{C}(z)$-differential automorphisms, $\sigma_1, \dots, \sigma_n$. The elementary symmetric functions of $\sigma_1(y_1), \dots, \sigma_n(y_1)$ are then invariant by G, hence belong to $\mathbb{C}(z)$ and y_1 is algebraic over $\mathbb{C}(z)$. Similarly, y_2 is algebraic over $\mathbb{C}(z)$; then L is a finite extension of $\mathbb{C}(z)$ and all solutions of (7.14) are algebraic over $\mathbb{C}(z)$.
4. Assume $G = \mathrm{SL}(2,\mathbb{C})$. If (7.14) had a Liouvillian solution, then all its solutions would be Liouvillian; hence $\mathbb{C}(z) \subset L$ would be a generalized Liouvillian extension. But $G^0 = \mathrm{SL}(2,\mathbb{C})$ is not solvable. (See Exercise 12 in chapter 4.) Hence $\mathbb{C}(z) \subset L$ is not Liouvillian (cf. Proposition 6.5.4).

□

Next Kovacic establishes necessary conditions for each of the three first cases in Theorem 7.3.1 to hold. These conditions give a sufficient condition for case 4 to hold, namely that all necessary conditions for cases 1, 2, and 3 fail.

Since r is a rational function, we may speak of the poles of r, by which we shall always mean the poles in the finite complex plane $\mathbb{C}$. If $r = s/t$, with $s, t \in \mathbb{C}[z]$ relatively prime, then the poles of r are the zeros of t and the order of the pole is the multiplicity of the zero of t. By the order of r at ∞, we shall mean the order of ∞ as a zero of r; thus the order of r at ∞ is $\deg t - \deg s$.

Theorem 7.3.2. *The following conditions are necessary for each of the respective cases in Theorem 7.3.1 to hold.*

Case 1. *Every pole of r must have even order or else have order 1; the order of r at ∞ must be even or else be greater than 2.*

Case 2. *r must have at least one pole that either has odd order greater than 2 or else has order 2.*

Case 3. *The order of a pole of r cannot exceed 2 and the order of r at ∞ must be at least 2. If the partial fraction expansion of r is*

$$r = \sum_i \frac{\alpha_i}{(z-c_i)^2} + \sum_j \frac{\beta_j}{z-d_j},$$

then $\sqrt{1+4\alpha_i} \in \mathbb{Q}$ for each i, $\sum_j \beta_j = 0$, and if $\gamma = \sum_i \alpha_i + \sum_j \beta_j d_j$, then $\sqrt{1+4\gamma} \in \mathbb{Q}$.

Proof. We examine the different cases.

Case 1. In this case, (7.14) has a solution of the form $y = e^{\int \omega}$, where $\omega \in \mathbb{C}(z)$. Since $y'' = ry$, ω satisfies the Riccati equation $\omega' + \omega^2 = r$. Both ω and r have Laurent series expansions around any point of the complex plane. To simplify notation, we consider the Laurent series expansion of ω and r at $z = 0$.

$$\omega = \sum_{m \geq \mu} a_m z^m, m \in \mathbb{Z}, a_m \in \mathbb{C}, a_\mu \neq 0,$$

$$r = \sum_{n \geq \nu} b_n z^n, n \in \mathbb{Z}, b_n \in \mathbb{C}, b_\nu \neq 0.$$

By substitution in the Riccati equation, we obtain

$$\mu a_\mu z^{\mu-1} + \cdots + a_\mu^2 z^{2\mu} + \cdots = b_\nu z^\nu + \ldots . \tag{7.15}$$

As we need to show that every pole of r either has order 1 or else has even order, we may assume that $\nu \leq -3$. Since $b_\nu \neq 0$, $-3 \geq \nu \geq \min(\mu - 1, 2\mu)$. It follows that $\mu < -1$ and $2\mu < \mu - 1$. Since $a_\mu^2 \neq 0$, we have $2\mu = \nu$ which implies that ν is even.

Now consider the Laurent series expansion of ω and r at ∞.

$$\omega = \sum_{m \leq \mu} a_m z^m, m \in \mathbb{Z}, a_m \in \mathbb{C}, a_\mu \neq 0,$$

$$r = \sum_{n \leq \nu} b_n z^n, n \in \mathbb{Z}, b_n \in \mathbb{C}, b_\nu \neq 0.$$

As we must show that either the order of r at ∞ is ≥ 3 or else is even, we may assume that $\nu \geq -1$. By substitution in the Riccati equation, we obtain

$$\mu a_\mu z^{\mu-1} + \cdots + a_\mu^2 z^{2\mu} + \cdots = b_\nu z^\nu + \ldots . \tag{7.16}$$

Just as above, $-1 \leq \nu \leq \max(\mu - 1, 2\mu), \mu > -1, 2\mu > \mu - 1$. Since $a_\mu^2 \neq 0$, $2\mu = \nu$, so ν is even. We have then obtained the necessary conditions for Case 1.

Case 2. In this case, by Theorem 7.3.1, the differential Galois group of (7.14) must be conjugate to a subgroup G of D^+, which is not triangularizable. (Otherwise case 1 would hold.) Let y_1, y_2 be a fundamental system of solutions of (7.14) relative to the group G. For every $\sigma \in G$, either $\sigma y_1 = c_\sigma y_1, \sigma y_2 = c_\sigma^{-1} y_2$ or $\sigma y_1 = -c_\sigma^{-1} y_2, \sigma y_2 = c_\sigma y_1$. Clearly $y_1^2 y_2^2$ is invariant by G, hence belongs to $\mathbb{C}(z)$. Moreover $y_1 y_2 \notin \mathbb{C}(z)$, for otherwise G would be a subgroup of the diagonal group, which is included in case 1.

Writing

$$y_1^2 y_2^2 = \prod (z - c_i)^{e_i} \ (e_i \in \mathbb{Z}),$$

we have that at least one exponent e_i is odd. Without loss of generality, we may assume that

$$y_1^2 y_2^2 = z^e \prod (z - c_i)^{e_i}$$

and that e is odd. Let

$$\theta = (y_1 y_2)'/(y_1 y_2) = \frac{1}{2}(y_1^2 y_2^2)'/(y_1^2 y_2^2) = \frac{1}{2} e z^{-1} + \ldots$$

where the dots represent terms of nonnegative order in z. Since $y_1'' = r y_1$ and $y_2'' = r y_2$, we have

$$\theta'' + 3\theta\theta' + \theta^3 = 4r\theta + 2r'.$$

Let $r = \sum_{n \geq \nu} b_n z^n$, where $n \in \mathbb{Z}, b_\nu \neq 0$, be the Laurent series expansion of r at 0. From the equation above, we obtain

$$(e - \frac{3}{4}e^2 + \frac{1}{8}e^3) z^{-3} + \cdots = 2b_\nu (e + \nu) z^{\nu-1} + \ldots .$$

If $\nu > -2$, then $0 = 8e - 6e^2 + e^3 = e(e-2)(e-4)$. This contradicts the fact that e is odd. Therefore $\nu \leq -2$. If $\nu < -2$, then $e + \nu = 0$, so ν is odd. We have then obtained the necessary conditions for case 2.

Case 3. In this case, (7.14) has a solution y which is algebraic over $\mathbb{C}(z)$, hence has a Puiseux series expansion around any point in the complex plane. We consider it at $z = 0$. Then $y = a_\mu z^\mu + \ldots$, where $\mu \in \mathbb{Q}$, $a_\mu \neq 0$

and the dots represent terms of order $> \mu$ in z. Let $r = \sum_{n \geq \nu} b_n z^n, n \in \mathbb{Z}, b_n \in \mathbb{C}, b_\nu \neq 0$, be the Laurent series expansion of r at 0. By substitution in (7.14), we obtain

$$\mu(\mu-1)a_\mu z^{\mu-2} + \cdots = a_\mu b_\nu z^{\nu+\mu} + \ldots.$$

It follows that $\nu \geq -2$, i.e. r has no pole of order greater than 2. If $\nu = -2$, then $\mu(\mu-1) = b_\nu$. As $\mu \in \mathbb{Q}$, we must have $\sqrt{1+4b_\nu} \in \mathbb{Q}$.

So far we have shown that the partial fraction expansion of r has the form

$$r = \sum_i \frac{\alpha_i}{(z-c_i)^2} + \sum_j \frac{\beta_j}{z-d_j} + P,$$

where $P \in \mathbb{C}[z]$ and $\sqrt{1+4\alpha_i} \in \mathbb{Q}$ for each i.

Next we consider the series expansions around ∞,

$$y = \sum_{m \leq \mu} a_m z^m, m \in \mathbb{Q}, a_m \in \mathbb{C}, a_\mu \neq 0,$$

$$r = \sum_{n \leq \nu} b_n z^n, n \in \mathbb{Z}, b_n \in \mathbb{C}, b_\nu \neq 0.$$

By substituting in (7.14), we obtain

$$\mu(\mu-1)a_\mu z^{\mu-2} + \cdots = \nu b_\nu a_\mu z^{\nu+\mu} + \ldots.$$

Just as above, we obtain $\nu \leq -2$ and therefore $P = 0$. But

$$r = \sum_i \frac{\alpha_i}{(z-c_i)^2} + \sum_j \frac{\beta_j}{z-d_j} = \Big(\sum_j \beta_j\Big) z^{-1} + \gamma z^{-2} + \ldots,$$

where $\gamma = \sum_i \alpha_i + \sum_j \beta_j d_j$. Therefore $\sum_j \beta_j = 0$ and $\mu(\mu-1) = \gamma$. Since $\mu \in \mathbb{Q}$, $\sqrt{1+4\gamma} \in \mathbb{Q}$. □

Example 7.3.3. 1. Consider the Airy equation

$$Y'' = zY.$$

As $r = z$ has order -1 at ∞ and no poles in the finite complex plane, we have that none of the cases 1, 2, 3 can hold. Hence the Airy equation has no Liouvillian solution.

2. Consider the Chebyshev differential equation

$$Y'' + \frac{z}{z^2-1}Y' - \frac{\alpha^2}{z^2-1}Y = 0, \alpha \in \mathbb{R}.$$

We can make a change of variable eliminating the first order term (see Exercise 1 in chapter 6) and obtain the equation

$$Y'' = \frac{(\alpha^2 - 1/4)z^2 - \alpha^2 - 1/2}{(z^2-1)^2}Y.$$

The coefficient r of Y in this equation is a function which has poles $1, -1$ of order 2 and its order at ∞ is 2, for $\alpha \neq \pm 1/2$. Hence cases 1, 2, and 4 can occur for any $\alpha \neq \pm 1/2$. The partial fraction expansion of r is

$$\frac{8\alpha^2+1}{16(z-1)} - \frac{8\alpha^2+1}{16(z+1)} - \frac{3}{16(z-1)^2} - \frac{3}{16(z+1)^2}.$$

Hence case 3 can occur if $\alpha \in \mathbb{Q} \setminus \{\pm 1/2\}$.

If $\alpha = \pm 1/2$, the order at ∞ is 4, and all four cases are possible.

3. Consider the differential equation $Y'' = rY$, where

$$r = \frac{101 - 81z^2}{48(1+3z^2)^2}.$$

The poles of r are $\pm i/\sqrt{3}$ of order 2 and the order of r at ∞ is 2; hence the necessary conditions for cases 1 and 2 are fulfilled.

The partial fraction expansion of r is

$$\frac{-37\sqrt{3}i}{288(z - i/\sqrt{3})} + \frac{37\sqrt{3}i}{288(z + i/\sqrt{3})} - \frac{2}{9(z - i/\sqrt{3})^2} - \frac{2}{9(z + i/\sqrt{3})^2}.$$

Hence the necessary conditions for case 3 are also fulfilled.

4. Consider the differential equation $Y'' = rY$, where

$$r = \frac{-128z^2 + 155z - 135}{576z^2(z-1)^2}.$$

The poles of r are $0, 1$ of order 2 and the order of r at infinity is 2, so the necessary conditions for cases 1 and 2 are fulfilled. The partial fraction expansion of r is

$$-\frac{115}{576z} + \frac{115}{576(z-1)} - \frac{15}{64z^2} - \frac{3}{16(z-1)^2}.$$

Hence the necessary conditions for case 3 are also fulfilled.

5. Consider the differential equation $Y'' = rY$, where

$$r = -\frac{3(25z^2 - (8/\sqrt{5})z + 19)}{16(5z^2 - 1)^2}.$$

The poles of r are $\pm 1/\sqrt{5}$ of order 2 and the order of r are ∞ is 2, so the necessary conditions for cases 1 and 2 are fulfilled. The partial fraction expansion of r is

$$\frac{21\sqrt{5}}{160(z - 1/\sqrt{5})} - \frac{21\sqrt{5}}{160(z + 1/\sqrt{5})} - \frac{21}{100(z - 1/\sqrt{5})^2} - \frac{6}{25(z + 1/\sqrt{5})^2}.$$

Hence the necessary conditions for case 3 are also fulfilled.

Once we have determined which of the cases given in Theorem 7.3.2 may occur for (7.14), we try to find a solution of the form given by Theorem 7.3.1. We now describe the algorithm for each of the cases.

7.3.2. The algorithm for case 1. The goal is to find a solution of (7.14) of the form $y = P\, e^{\int \vartheta}$, where $P \in \mathbb{C}[z]$ and $\vartheta \in \mathbb{C}(z)$. Since $P\, e^{\int \vartheta} = e^{\int (\frac{P'}{P} + \vartheta)}$, this is of the form given in Theorem 7.3.1 for case 1, with $\omega = \frac{P'}{P} + \vartheta$. If the partial fraction expansion of a rational function g is

$$g = Q + \sum_c \sum_{j=1}^{\nu_c} \frac{a_{cj}}{(z - c)^j}, \tag{7.17}$$

with $Q \in \mathbb{C}[z]$, then

$$\sum_{j=1}^{\nu_c} \frac{a_{cj}}{(z - c)^j}$$

is the sum of the negative degree terms of the Laurent series expansion of g at the pole c. We shall refer to it as the partial fraction expansion of g at c. The sum of the nonnegative degree terms of the Laurent series expansion of g at ∞ is equal to the polynomial Q in (7.17). The local conditions given by Theorem 7.3.2 allow us to determine the partial fraction expansion of ω at each of the poles of r and at infinity. Then these expansions are "glued" together to form a candidate for ϑ. The polynomial P will take account of the poles of ω which are not poles of r.

If $g = \sum_{m \geq \mu} a_m (z - c)^m, m \in \mathbb{Z}$, is the Laurent series expansion of a meromorphic function g at a pole c, we shall use, following Kovacic, the notation

$$(7.18)\qquad \begin{array}{rcl} [g]_c &=& \sum_{\mu\le m\le -2} a_m(z-c)^m, \\ \alpha_c &=& a_{-1} \text{ (the residue of } g \text{ at } c), \\ \overline{g}_c &=& \sum_{m\ge 0} a_m(z-c)^m. \end{array}$$

If $g=\sum_{m\le\mu} a_m z^m$, $m\in\mathbb{Z}$, is the Laurent series expansion of a meromorphic function g at ∞, we set

$$(7.19)\qquad \begin{array}{rcl} [g]_\infty &=& \sum_{0\le m\le \mu} a_m z^m, \\ \alpha_\infty &=& a_{-1}, \\ \overline{g}_\infty &=& \sum_{m\le -2} a_m z^m. \end{array}$$

Let us observe that $-a_{-1}$ is the residue at ∞ of the differential form gdz.

Let Γ denote the set of poles of r in the complex plane. The first step of the algorithm will be determining the possible values for $[\omega]_c$ and $\mathrm{Res}(\omega,c)$ for each $c\in\Gamma\cup\{\infty\}$. The second step discards some of the combinations of local data by using the relation between the residues of a meromorphic function at its different poles. For the remaining possibilities, the third step is trying to find a suitable polynomial P.

Proposition 7.3.4. *Let $r\in\mathbb{C}(z)$ satisfy the necessary conditions for case 1 given in Theorem 7.3.2. Let Γ denote the set of poles of r in the complex plane.*

Step 1. For each $c\in\Gamma\cup\{\infty\}$, we define a rational function $[\omega]_c$ and two complex numbers α_c^+,α_c^- as described below.

(c1) If $c\in\Gamma$ and c is a pole of order 1, then

$$[\omega]_c=0,\quad \alpha_c^+=\alpha_c^-=1.$$

(c2) If $c\in\Gamma$ and c is a pole of order 2, then

$$[\omega]_c=0,\quad \alpha_c^\pm=\frac{1}{2}\pm\frac{1}{2}\sqrt{1+4b},$$

for b the coefficient of $(z-c)^{-2}$ in the partial fraction expansion for r.

(c3) If $c\in\Gamma$ and c is a pole of order $2\nu\ge 4$, then

$$[\omega]_c=\pm[\sqrt{r}]_c,\quad \alpha_c^\pm=\frac{1}{2}\left(\pm\frac{b}{a}+\nu\right),$$

where a is the coefficient of $(z-c)^{-\nu}$ in $[\sqrt{r}]_c$ and b is the coefficient of $(z-c)^{-\nu-1}$ in $r-[\sqrt{r}]_c^2$.

(∞1) *If the order of r at ∞ is > 2, then*

$$[\omega]_\infty = 0, \quad \alpha_\infty^+ = 0, \quad \alpha_\infty^- = 1.$$

(∞2) *If the order of r at ∞ is 2, then*

$$[\omega]_\infty = 0, \quad \alpha_\infty^\pm = \frac{1}{2} \pm \frac{1}{2}\sqrt{1+4b},$$

where b is the coefficient of $1/z^2$ in the Laurent series expansion of r at ∞.

(∞3) *If the order of r at ∞ is $-2\nu \leq 0$, then*

$$[\omega]_\infty = \pm[\sqrt{r}]_\infty, \quad \alpha_\infty^\pm = \frac{1}{2}\left(\pm\frac{b}{a} - \nu\right),$$

where a is the coefficient of z^ν in the Laurent series expansion of $\sqrt{r}$ at ∞ and b is the coefficient of $z^{\nu-1}$ in $r - [\sqrt{r}]_\infty^2$.

Step 2. For each family $s = (s(c))_{c\in\Gamma\cup\{\infty\}}$, where $s(c) = +$ or $-$, let

$$d_s = \alpha_\infty^{s(\infty)} - \sum_{c\in\Gamma} \alpha_c^{s(c)}.$$

If d_s is a nonnegative integer, let

$$\vartheta_s = \sum_{c\in\Gamma}\left(s(c)[\omega]_c + \frac{\alpha_c^{s(c)}}{z-c}\right) + s(\infty)[\omega]_\infty.$$

Step 3. For each of the ϑ_s considered in step 2, search for a polynomial P_s of degree d_s satisfying the differential equation

$$P_s'' + 2\vartheta_s P_s' + (\vartheta_s' + \vartheta_s^2 - r)P_s = 0. \tag{7.20}$$

If such a polynomial exists for some s, then $y = Pe^{\int\vartheta}$ is a solution of (7.14) with $P = P_s, \vartheta = \vartheta_s$.

If d_s is not a nonnegative integer for any of the families s considered in step 2 or there is no polynomial solution to (7.20) for any of the families s retained for step 3, then (7.14) has no solution of the form $y = e^{\int\omega}$, with $\omega \in \mathbb{C}(z)$.

Proof. We shall first prove that, if $y = e^{\int\omega}$ is a solution to (7.14), then the partial fraction expansion $\sum_{i=1}^\nu a_i/(z-c)^i$ of ω at c for $c \in \Gamma$ (resp. $c = \infty$), has the form described in step 1 of the statement for each case. If (7.14) has a solution of the form $y = e^{\int\omega}$, then ω satisfies the Riccati equation, i.e.

$$\omega' + \omega^2 = r.$$

Let $c \in \Gamma$. By a change of variable, we may assume that $c = 0$ and drop the subscript.

(c1) Suppose that 0 is a pole of r of order 1 , then $r = b/z + \dots$, where $b \neq 0$. The Riccati equation becomes

$$-\frac{\nu a_\nu}{z^{\nu+1}} + \cdots + \frac{a_\nu^2}{z^{2\nu}} + \cdots = \frac{b}{z} + \cdots .$$

Since $a_\nu \neq 0$, $\nu \leq 1$ and $[\omega] = 0$. Substituting $\omega = \alpha/z + \overline{\omega}$ into the Riccati equation, we have

$$-\frac{\alpha}{z^2} + \overline{\omega} + \frac{\alpha^2}{z^2} + \frac{2\alpha}{z}\overline{\omega} + \overline{\omega}^2 = \frac{b}{z} + \cdots .$$

Therefore $-\alpha + \alpha^2 = 0$, so $\alpha = 0$ or $\alpha = 1$. But, for $\alpha = 0$, the point 0 would be an ordinary point for the left-hand side of this equation, so $\alpha = 1$.

(c2) If 0 is a pole of r of order 2, then $r = b/z^2 + \dots$, where $b \neq 0$. As in (c1), $[\omega] = 0$ and $-\alpha + \alpha^2 = b$. Thus the partial fraction expansion of ω at 0 is

$$\frac{\alpha^\pm}{z}, \text{ where } \alpha^\pm = \frac{1}{2} \pm \frac{1}{2}\sqrt{1 + 4b}.$$

(c3) If 0 is a pole of r of order $2\nu \geq 4$, then 0 is a pole of ω of order ν, as we saw in the proof of Theorem 7.3.2 from (7.15). We put $\widetilde{r} = \sqrt{r} - [\sqrt{r}]$. We then have $r = [\sqrt{r}]^2 + 2\widetilde{r}[\sqrt{r}] + \widetilde{r}^2$. This equality, together with $\omega = [\omega] + \dfrac{\alpha}{z} + \overline{\omega}$ and the Riccati equation, gives

$$\begin{aligned}([\omega] + [\sqrt{r}]) \cdot ([\omega] - [\sqrt{r}]) = -[\omega]' + \frac{\alpha^2}{z^2} - \overline{\omega}' - \frac{2\alpha}{z}[\omega] \\ -2\overline{\omega}[\omega] - \frac{\alpha^2}{z^2} - \frac{2\alpha}{z}\overline{\omega} - \overline{\omega}^2 + 2\widetilde{r}[\sqrt{r}] + \widetilde{r}^2.\end{aligned} \tag{7.21}$$

We may observe that the right-hand side of this equation does not have terms in z^{-i}, for $i \geq \nu + 2$ as $\nu \geq 1$. Now $([\omega] + [\sqrt{r}) + ([\omega] - [\sqrt{r}) = 2[\omega]$. Hence at least one of the factors in the left-hand side of the equation has a term in $z^{-\nu}$. If the other factor were not zero, it would have some term in z^{-i}, for $i \geq 2$, which would contradict the observation above. We then obtain $[\omega] = \pm[\sqrt{r}]$.

The coefficient of $z^{-\nu-1}$ in the right-hand side of (7.21) is $\pm\nu a \mp 2\alpha a + b$, where a is the coefficient of $z^{-\nu}$ in $[\sqrt{r}]$ and b is the coefficient of $z^{-\nu-1}$ in $2\widetilde{r}[\sqrt{r}] + \widetilde{r}^2 = r - [\sqrt{r}]^2$. Therefore $\alpha^\pm = \dfrac{1}{2}(\pm b/a + \nu)$.

(c4) Finally we consider what happens when 0 is an ordinary point of r. As in (c1), $[\omega] = 0$ and $-\alpha + \alpha^2 = 0$. Contrary to the situation in (c1), we cannot conclude that $\alpha \neq 0$. Hence the component of the partial fraction expansion of ω at 0 is either 0 or $1/z$.

We collect together what we have proved so far. Let Γ be the set of poles of r. Then

$$\omega = \sum_{c\in\Gamma}\left(s(c)[\omega]_c + \frac{\alpha_c^{s(c)}}{z-c}\right) + \sum_{i=1}^{n}\frac{1}{z-d_i} + R,$$

where $R \in \mathbb{C}[z], s(c) = +$ or $-$ and $[\omega]_c, \alpha_s^{s(c)}$ are as in the statement of the proposition. We now consider the Laurent series expansion of ω at ∞. Then R will be determined by $[\omega]_\infty$.

(∞1) If r has order $\nu > 2$ at ∞, then $r = \sum_{i\geq\nu}\frac{b_\nu}{z^\nu}$. The Riccati equation implies that $[\omega]_\infty = 0$ and $-\alpha_\infty + \alpha_\infty^2 = 0$, so $\alpha_\infty = 0$ or 1.

(∞2) If r has order 2 at ∞, then $r = \sum_{i\geq 2}\frac{b_\nu}{z^\nu}$. The Riccati equation implies that $[\omega]_\infty = 0$ and $-\alpha_\infty + \alpha_\infty^2 = b_2$, so $\alpha_\infty = \frac{1}{2} \pm \frac{1}{2}\sqrt{1+4b_2}$.

(∞3) In the other cases, the order of r at ∞ must be even, by the necessary conditions in Theorem 7.3.2. Reasoning as we did in (c3), we find that $[\omega]_\infty = \pm[\sqrt{r}]_\infty, \alpha_\infty = \frac{1}{2}\left(\pm\frac{b}{a} - \nu\right)$, where a is the coefficient of z^ν in $[\sqrt{z}]_\infty$ and b is the coefficient of $z^{\nu-1}$ in $r - [\sqrt{r}]_\infty^2$.

We now know that the partial fraction expansion of ω has the form

$$\omega = \sum_{c\in\Gamma}\left(s(c)[\omega]_c + \frac{\alpha_c^{s(c)}}{z-c}\right) + \sum_{i=1}^{n}\frac{1}{z-d_i} + s(\infty)[\omega]_\infty.$$

Now we apply the residue theorem (see e.g. [**F**], 10.21) to the 1-form ωdz. Its residue at a point c of the complex plane is α_c and its residue at ∞ is $-\alpha_\infty$. We then obtain $\sum_{c\in\Gamma}\alpha_c + n - \alpha_\infty = 0$; hence $\alpha_\infty - \sum_{c\in\Gamma}\alpha_c \in \mathbb{N}$.

Let

$$\vartheta = \sum_{c\in\Gamma}\left(s(c)[\omega]_c + \frac{\alpha_c^{s(c)}}{z-c}\right) + s(\infty)[\omega]_\infty$$

and

$$P = \prod_{i=1}^{n}(z-d_i).$$

Then $\omega = \vartheta + P'/P$. Again, using the Riccati equation, we obtain

$$P'' + 2\vartheta P' + (\vartheta' + \vartheta^2 - r)P = 0.$$

Conversely, if P satisfies this equation, then $\omega = \vartheta + P'/P$ satisfies the Riccati equation. Hence $y = Pe^{\int \vartheta}$ is a solution of the differential equation (7.14). □

Example 7.3.5. 1. Let us again consider the Chebyshev differential equation

$$Y'' = \frac{(\alpha^2 - 1/4)z^2 - \alpha^2 - 1/2}{(z^2-1)^2} Y. \tag{7.22}$$

We shall now determine if it admits a solution of the form $y = e^{\int \omega}$, with $\omega \in \mathbb{C}(z)$.

We have

$$[\omega]_1 = [\omega]_{-1} = 0, \alpha_1^+ = \alpha_{-1}^+ = \frac{3}{4}, \alpha_1^- = \alpha_{-1}^- = \frac{1}{4}.$$

For $\alpha \neq \pm 1/2$,

$$[\omega]_\infty = 0, \alpha_\infty^\pm = \frac{1}{2} \pm \frac{i}{2\sqrt{2}}.$$

For $\alpha = \pm 1/2$,

$$[\omega]_\infty = 0, \alpha_\infty^+ = 0, \alpha_\infty^- = 1.$$

Hence $d = \alpha_\infty^{s(\infty)} - \alpha_1^{s(1)} - \alpha_{-1}^{s(-1)}$ is a nonnegative integer only when $\alpha = \pm 1/2$ for $s(-1) = -s(1), s(\infty) = -$ and in this case d is 0. So the differential equation (7.22) has no solution of the form $e^{\int \omega}$, with $\omega \in \mathbb{C}(z)$ for $\alpha \neq \pm 1/2$. The candidates for ϑ in the case $\alpha = \pm 1/2$ are

$$\vartheta_1 = \frac{3/4}{z-1} + \frac{1/4}{z+1}, \quad \vartheta_2 = \frac{1/4}{z-1} + \frac{3/4}{z+1}.$$

As $d = 0$, we just need to check if ϑ satisfies the Riccati equation

$$\vartheta' + \vartheta^2 = \frac{-3/4}{(z^2-1)^2}.$$

We obtain that both candidates for ϑ do, so the differential equation

$$Y'' = \frac{-3}{4(z^2-1)^2} Y$$

admits the independent Liouvillian solutions

$$y_1 = (z-1)^{3/4}(z+1)^{1/4}, \quad y_2 = (z-1)^{1/4}(z+1)^{3/4}.$$

We observe that both are algebraic over $\mathbb{C}(z)$ and satisfy $y_1y_2 \in \mathbb{C}(z)$. The differential Galois group of the differential equation is the cyclic group C_4.

2. Let us again consider the equation in Example 7.3.3.**3**. The poles of r are $c_1 = i/\sqrt{3}, c_2 = -i/\sqrt{3}$ of order 2. We then have, following the notations in Proposition 7.3.4, $[\omega]_{c_1} = [\omega]_{c_2} = 0, \alpha_{c_1}^+ = \alpha_{c_2}^+ = 2/3, \alpha_{c_1}^- = \alpha_{c_2}^- = 1/3$; $[\omega]_\infty = 0, \alpha_\infty^+ = 5/8, \alpha_\infty^- = 3/8$. Then $d_s = \alpha_\infty^{s(\infty)} - \alpha_{c_1}^{s(c_1)} - \alpha_{c_2}^{s(c_2)}$ is never an integer. Hence case 1 is not possible.

3. Let us again consider the equation in Example 7.3.3.**4**. The poles of r are $0, 1$ of order 2. We then have $[\omega]_0 = [\omega]_1 = 0, \alpha_0^+ = 5/8, \alpha_0^- = 3/8, \alpha_1^+ = 1, \alpha_1^- = 0$; $[\omega]_\infty = 0, \alpha_\infty^+ = 2/3, \alpha_\infty^- = 1/3$. Then $d_s = \alpha_\infty^{s(\infty)} - \alpha_0^{s(0)} - \alpha_1^{s(1)}$ is never an integer. Hence case 1 is not possible.

4. Let us again consider the equation in Example 7.3.3.**5**. The poles of r are $c_1 = 1/\sqrt{5}, c_2 = -1/\sqrt{5}$ of order 2. We then have $[\omega]_{c_1} = [\omega]_{c_2} = 0, \alpha_{c_1}^+ = 7/10, \alpha_{c_1}^- = 3/10, \alpha_{c_2}^+ = 3/5, \alpha_{c_2}^- = 2/5$; $[\omega]_\infty = 0, \alpha_\infty^+ = 3/4, \alpha_\infty^- = 1/4$. Then $d_s = \alpha_\infty^{s(\infty)} - \alpha_{c_1}^{s(c_1)} - \alpha_{c_2}^{s(c_2)}$ is never an integer. Hence case 1 is not possible.

7.3.3. The algorithm for case 2. The goal is to find a solution of (7.14) of the form $y = e^{\int \omega}$, where ω is quadratic over $\mathbb{C}(z)$. As in case 1, we will collect local data at the poles of r and at ∞. From these, some will be retained and used to form candidates for a rational function ϑ. For each of these candidates, we shall search for a polynomial $P \in \mathbb{C}[z]$ of a certain degree satisfying a differential equation. If no such polynomial exists for any family, then case 2 cannot hold. If such a polynomial does exist, then ω will be obtained as a root of a quadratic equation whose coefficients are given in terms of the rational function $\phi = \vartheta + \frac{P'}{P}$.

Proposition 7.3.6. *Let $r \in \mathbb{C}(z)$ satisfy the necessary conditions for case 2 given in Theorem 7.3.2. Let Γ denote the set of poles of r in the complex plane.*

Step 1. For each $c \in \Gamma \cup \{\infty\}$, we define a set E_c as described below.
(c1) If $c \in \Gamma$ and c is a pole of order 1, then

$$E_c = \{4\}.$$

(c2) If $c \in \Gamma$ and c is a pole of order 2, then

$$E_c = \{2 + k\sqrt{1+4b} \, : \, k = 0, \pm 2\} \cap \mathbb{Z},$$

for b the coefficient of $(z-c)^{-2}$ in the partial fraction expansion for r.

(c3) If $c \in \Gamma$ and c is a pole of order $\nu > 2$, then

$$E_c = \{\nu\}.$$

(∞1) If the order of r at ∞ is > 2, then

$$E_\infty = \{0, 2, 4\}.$$

(∞2) If the order of r at ∞ is 2, then

$$E_\infty = \{2 + k\sqrt{1+4b} \,:\, k = 0, \pm 2\} \cap \mathbb{Z},$$

where b is the coefficient of $1/z^2$ in the Laurent series expansion of r at ∞.

(∞3) If the order of r at ∞ is $\nu < 2$, then

$$E_\infty = \{\nu\}.$$

Step 2. We consider the families $(e_c)_{c\in\Gamma\cup\infty}$ with $e_c \in E_c$. Those families, all of whose coordinates are even, may be discarded. For each remaining family, let

$$d = \frac{1}{2}(e_\infty - \sum_{c\in\Gamma} e_c).$$

If d is a nonnegative integer, let

$$\vartheta = \frac{1}{2}\sum_{c\in\Gamma} \frac{e_c}{z-c}.$$

Step 3. For each of the ϑ considered in step 2, search for a polynomial P of degree d satisfying the differential equation

$$\begin{aligned} P''' + \vartheta P'' + (3\vartheta^2 + 3\vartheta' - 4r)P' & \\ +(\vartheta'' + 3\vartheta\vartheta' + \vartheta^3 - 4r\vartheta - 2r')P &= 0. \end{aligned} \tag{7.23}$$

If such a polynomial exists for some (e_c), then set $\phi = \vartheta + P'/P$ and let ω be a solution of the quadratic equation

$$\omega^2 + \phi\omega + (\frac{1}{2}\phi' + \frac{1}{2}\phi^2 - r) = 0. \tag{7.24}$$

Then $y = e^{\int \omega}$ is a solution to (7.14).

If d is not a nonnegative integer for any of the families (e_c) considered in step 2 or there is no polynomial solution to (7.23) for any of the families (e_c) retained for step 3, then (7.14) has no solution of the form $y = e^{\int \omega}$, with ω quadratic over $\mathbb{C}(z)$.

Proof. In case 2, the differential Galois group G of the differential equation (7.14) is conjugate to a subgroup of

$$D^+ = \left\{ \begin{pmatrix} c & 0 \\ 0 & c^{-1} \end{pmatrix} : c \in C, c \neq 0 \right\} \cup \left\{ \begin{pmatrix} 0 & c \\ -c^{-1} & 0 \end{pmatrix} : c \in C, c \neq 0 \right\}$$

which is not triangularizable as case 1 does not hold. Let y_1, y_2 be a fundamental set of solutions of (7.14) corresponding to the subgroup of D^+. For any differential automorphism σ of $\mathbb{C}(z)\langle y_1, y_2\rangle$ over $\mathbb{C}(z)$, either $\sigma y_1 = c y_1, \sigma y_2 = c^{-1} y_2$ or $\sigma y_1 = -c^{-1} y_2, \sigma y_2 = c y_1$, for some $c \in \mathbb{C}, c \neq 0$. We then have $\sigma(y_1^2 y_2^2) = y_1^2 y_2^2$, for all σ; hence $y_1^2 y_2^2 \in \mathbb{C}(z)$. Moreover, $y_1 y_2 \notin \mathbb{C}(z)$, as G is not triangularizable and so $\sigma(y_1 y_2) = -y_1 y_2$, for some $\sigma \in G$. We write

$$(y_1 y_2)^2 = \prod_{c \in \Gamma} (z - c)^{e_c} \prod_{i=1}^{m} (z - d_i)^{f_i},$$

where Γ is the set of poles of r and the exponents e_c, f_i are integers. Our goal is to determine these exponents. Let

$$\phi = \frac{(y_1 y_2)'}{y_1 y_2} = \frac{(y_1^2 y_2^2)'}{2 y_1^2 y_2^2} = \frac{1}{2} \sum_{c \in \Gamma} \frac{e_c}{z - c} + \frac{1}{2} \sum_{i=1}^{m} \frac{f_i}{z - d_i}.$$

As $\phi = y_1'/y_1 + y_2'/y_2$, it follows that

$$\phi'' + 3\phi\phi' + \phi^3 = 4r\phi + 2r'. \tag{7.25}$$

We first determine e_c for $c \in \Gamma$. In order to simplify the notation, we assume that $c = 0$.

(c1) Let us assume that 0 is a pole of r of order 1. The Laurent series expansions of r and ϕ are of the form

$$r = b_{-1} z^{-1} + \ldots \ (b_{-1} \neq 0)$$
$$\phi = \frac{1}{2} e z^{-1} + f + \ldots \ (e \in \mathbb{Z}, f \in \mathbb{C}).$$

Substituting these series in (7.25) and retaining all those terms that involve z^{-3} and z^{-2}, we obtain the following

$$ez^{-3}+\cdots-\frac{3}{4}e^2z^{-3}-\frac{3}{2}efz^{-2}+\cdots+\frac{1}{8}e^3z^{-3}+\frac{3}{4}fz^{-2}+\ldots$$
$$=b_{-1}ez^{-2}+\cdots-2b_{-1}z^{-2}+\cdots.$$

Therefore $e-\frac{3}{4}e^2+\frac{1}{8}e^3=0$, so $e=0,2,4$. Also $-\frac{3}{2}ef+\frac{3}{4}e^2f=2b_{-1}e-2b_{-1}$. As $b_{-1}\neq 0$, $e\neq 0,2$. Hence e must be 4.

(c2) Assume that 0 is a pole of r of order 2 and let

$$r=b_{-2}z^{-2}+\ldots(b_{-2}\neq 0)$$
$$\phi=\frac{1}{2}ez^{-1}+\ldots.$$

Equating the coefficients of z^{-3} on the two sides of (7.25), we obtain

$$e-\frac{3}{4}e^2+\frac{1}{8}e^3=2eb_{-2}-4b_{-2}.$$

The roots of this equation are $e=2, e=2\pm2\sqrt{1+4b_{-2}}$. Of course, the latter two roots may be discarded in the case that they are nonintegral.

(c3) Assume that 0 is a pole of r of order $\nu>2$. Then

$$r=b_{-\nu}z^{-\nu}+\ldots(b_{-\nu}\neq 0)$$
$$\phi=\frac{1}{2}ez^{-1}+\ldots.$$

Equating the coefficients of $z^{-\nu-1}$ on the two sides of (7.25), we obtain

$$0=2eb_{-\nu}-2\nu b_{-\nu}.$$

Hence $e=\nu$.

(c4) Finally, if 0 is an ordinary point of r, as in (c1), we obtain $f_i=0,2,4$. We can of course exclude the possibility $f_i=0$.

We have shown so far that

$$y_1^2y_2^2=\prod_{c\in\Gamma}(z-c)^{e_c}P^2,$$

where $e_c\in E_c$ and $P\in\mathbb{C}[z]$. Set

$$\vartheta=\frac{1}{2}\sum_{c\in\Gamma}\frac{e_c}{z-c},$$

so

$$\phi=\frac{1}{2}\frac{(y_1^2y_2^2)'}{y_1^2y_2^2}=\vartheta+\frac{P'}{P}.$$

The next step of the proof is to determine the degree d of P. By the residue theorem (applied to the differential 1-form ϕdz), we obtain $e_\infty = \sum_{c\in\Gamma} e_c + 2d$, for e_∞ the coefficient of z in the Laurent series expansion of ϕ at ∞. We discuss the different cases according to the order of r at ∞.

(∞1) Assume that the order of r at ∞ is 1. As in (c1) we find that $e_\infty = 0, 2$ or 4.

(∞2) Assume that the order of r at ∞ is 2. As in (c2) we find that $e_\infty = 2, 2 \pm 2\sqrt{1+4b_{-2}}$, where b_{-2} is the coefficient of z^{-2} in the Laurent series expansion of ϕ at ∞ and e_∞ must be integral.

(∞3) Assume that the order of r at ∞ is ν. As in (c3) we find that $e_\infty = \nu$.

Note that at least one of the $e_c (c \in \Gamma)$ is odd since $y_1 y_2 \notin \mathbb{C}(z)$.

Using (7.25) and the equation $\phi = \vartheta + P'/P$, we obtain

(7.26) $$P''' + 3\vartheta P'' + (3\theta^2 + 3\theta' - 4r)P' + (\vartheta'' + 3\vartheta\vartheta' + \vartheta^3 - 4r\theta - 2r')P = 0.$$

Now we have $\phi = \dfrac{y_1'}{y_1} + \dfrac{y_2'}{y_2}$ and $\phi' = 2r - \dfrac{y_1'^2}{y_1^2} - \dfrac{y_2'^2}{y_2^2} = 2r - \phi^2 + 2\dfrac{y_1'}{y_1}.\dfrac{y_2'}{y_2}$.
Hence $\dfrac{y_1'}{y_1}.\dfrac{y_2'}{y_2} = \dfrac{1}{2}\phi' + \dfrac{1}{2}\phi^2 - r$. Let ω be a root of the quadratic equation

(7.27) $$\omega^2 - \phi\omega + \frac{1}{2}\phi' + \frac{1}{2}\phi^2 - r = 0.$$

To complete the proof we need to show that $y = e^{\int \omega}$ is a solution of the differential equation (7.14). From (7.27), we obtain by differentiation

$$(2\omega - \phi)\omega' = \phi'\omega - \frac{1}{2}\phi'' - \phi\phi' + r'.$$

The factor $(2\omega - \phi)$ cannot be zero. Indeed, if $\phi = 2\omega$, then (7.27) would give $\omega' + \omega^2 - r = 0$, so $y = e^{\int \omega}$ would be a solution of (7.14). But then $\omega = \dfrac{1}{2}\phi \in \mathbb{C}(z)$, which is excluded, as we are assuming case 1 fails. Using (7.27) and (7.25), we have

$$2(2\omega - \phi)(\omega' + \omega^2 - r) = -\phi'' - 3\phi\phi' - \phi^3 + 4r\phi + 2r' = 0.$$

Thus $\omega' + \omega^2 = r$ so $y = e^{\int \omega}$ is a solution of the differential equation (7.14). □

Example 7.3.7. 1. Let us again consider the Chebyshev differential equation

$$(7.28) \qquad Y'' = \frac{(\alpha^2 - 1/4)z^2 - \alpha^2 - 1/2}{(z^2-1)^2} Y,$$

with $\alpha \neq \pm 1/2$. We shall now determine if it admits a solution of the form $y = e^{\int \omega}$, with ω quadratic over $\mathbb{C}(z)$. We have

$$E_1 = E_{-1} = \{1,2,3\}, \quad E_\infty = 2.$$

The unique triple (e_1, e_{-1}, e_∞), not having all components even, satisfying that $d = (e_\infty - e_1 - e_{-1})/2$ is a nonnegative integer is $(1,1,2)$ and gives $d = 0$. We then have

$$\vartheta = \frac{1}{2}\left(\frac{1}{z-1} + \frac{1}{z+1}\right).$$

As d=0, we look for ω satisfying the quadratic equation

$$\omega^2 + \vartheta\omega + (\frac{1}{2}\vartheta' + \frac{1}{2}\vartheta^2 - r) = 0.$$

The root

$$\omega = \frac{z + 2\alpha\sqrt{z^2-1}}{2(z^2-1)}$$

satisfies the Riccati equation. Hence

$$y = e^{\int \omega} = (z^2-1)^{1/4}(z + \sqrt{z^2-1})^\alpha$$

is a solution to the Chebyshev differential equation.

2. Let us again consider the equation in Example 7.3.3.**3**. The poles of r are $c_1 = i/\sqrt{3}, c_2 = -i/\sqrt{3}$ of order 2. Following the notations in Proposition 7.3.6, we obtain $E_{c_1} = E_{c_2} = E_\infty = \{2\}$, so the only possible value of $d = (e_\infty - e_{c_1} - e_{c_2})/2$ is negative and case 2 is not possible.

3. Let us again consider the equation in Example 7.3.3.**4**. The poles of r are $0, 1$ of order 2. We obtain $E_0 = \{2\}, E_1 = \{1,2,3\}, E_\infty = \{2\}$, so $d = (e_\infty - e_{c_1} - e_{c_2})/2$ is always negative and case 2 is not possible.

4. Let us again consider the equation in Example 7.3.3.**5**. The poles of r are $c_1 = 1/\sqrt{5}, c_2 = -1/\sqrt{5}$ of order 2. We have $E_{c_1} = E_{c_2} = \{2\}, E_\infty = \{1,2,3\}$, so $d = (e_\infty - e_{c_1} - e_{c_2})/2$ is always negative and case 2 is not possible.

7.3.4. The algorithm for case 3. In this section, we describe the algorithm for case 3. Taking into account Theorem 4.6.1 and Proposition 4.6.10, in this case the differential Galois group of the differential equation is either the tetrahedral group, the octahedral group, or the icosahedral group. The aim of the algorithm is to find an irreducible polynomial $A \in \mathbb{C}(z)[T]$ such that if ω is a root of A, then $y = e^{\int \omega}$ is a solution to (7.14).

The next proposition indicates how we can find the coefficients of the polynomial A. We recall that $y = e^{\int \omega}$ being a solution to $Y'' = rY$ is equivalent to ω being a solution to the Riccati equation $\omega' + \omega^2 = r$.

Proposition 7.3.8. *Let ω satisfy $\omega' + \omega^2 = r$ and let*

$$T^n - \sum_{i=0}^{n-1} \frac{a_i}{(n-i)!} T^i$$

be the minimal polynomial of ω over $\mathbb{C}(z)$. Then the coefficients a_i satisfy

$$(n-i)(i+1)ra_{i+1} + a_{i-1} + a_i' + sa_i = 0,$$

where $s = a_{n-1}$ and we put $a_n = -1, a_{-1} = 0$.

Proof. Put $A = \sum_{i=0}^{n} \frac{a_i}{(n-i)!} T^i$, with $a_n = -1$. Consider the polynomial

$$B = \frac{\partial A}{\partial T}(r - T^2) + \frac{\partial A}{\partial z} + (nT + s)A,$$

where $s = a_{n-1}$. The coefficient of T^{n+1} in B is $-na_n + na_n = 0$ and the coefficient of T^n in B is

$$-(n-1)a_{n-1} + a_n' + na_{n-1} + sa_n = a_{n-1} - s = 0,$$

since $a_n = -1$ and $s = a_{n-1}$. Therefore B has degree $< n$ in T. But

$$B(\omega) = \frac{\partial A}{\partial T}(\omega)(r - \omega^2) + \frac{\partial A}{\partial z}(\omega) + (n\omega + s)A(\omega) = \frac{d}{dz}(A(\omega)) + (n\omega + s)A(\omega) = 0.$$

Hence $B = 0$. The coefficient of T^i in B is

$$\begin{aligned} 0 &= (i+1)\frac{a_{i+1}}{(n-1-i)!}r - (i-1)\frac{a_{i-1}}{(n-1-i)!} + \frac{a_i'}{(n-i)!} \\ &\quad + n\frac{a_{i-1}}{(n-1-i)!} + s\frac{a_i}{(n-i)!} \\ &= \frac{1}{(n-i)!}[(n-i)(i+1)ra_{i+1} + a_{i-1} + a_i' + sa_i], \end{aligned}$$

where $a_{-1} = 0$, which gives the relations in the statement. □

Our aim is now to establish a converse of the preceding proposition. To this end, we consider the following recursive differential equation

$$
\text{(7.29)} \qquad \begin{array}{rcl} a_n & = & -1 \\ a_{i-1} & = & -a_i' - sa_i - (n-i)(i+1)ra_{i+1},\ i = n, \dots, 0 \end{array}
$$

and define a solution of (7.29) as an element s in $\mathbb{C}(z)$ such that when $a_n, \dots, a_{-1}$ are defined as above, then a_{-1} is identically 0.

Proposition 7.3.9. *Let s be a solution of (7.29) for some n and let ω be any root of the polynomial*

$$
A = \sum_{i=0}^{n} \frac{a_i}{(n-i)!} T^i .
$$

Then $y = e^{\int \omega}$ is a solution of the differential equation $Y'' = rY$.

Proof. We claim that

$$
\frac{\partial^{k+1}A}{\partial T^{k+1}}(T^2-r) = \frac{\partial^{k+1}A}{\partial T^k \partial z} + ((n-2k)T+s)\frac{\partial^k A}{\partial T^k} + k(n-k+1)\frac{\partial^{k-1}A}{\partial T^{k-1}},\ k = 0, 1, \dots .
$$

For $k = 0$, we have

$$
\begin{aligned}
\frac{\partial A}{\partial T}(T^2 - r) &= \left(\textstyle\sum_{i=1}^{n} \frac{ia_i}{(n-i)!} T^{i-1}\right)(T^2 - r) \\
&= na_n T^{n+1} + \textstyle\sum_{i=0}^{n-1} \frac{ia_i}{(n-i)!} T^{i+1} - \sum_{i=0}^{n-1} \frac{(i+1)ra_{i+1}}{(n-1-i)!} T^i \\
&= nTA - \textstyle\sum_{i=0}^{n-1} \frac{(n-i)a_i}{(n-i)!} T^{i+1} - \sum_{i=0}^{n-1} \frac{(n-i)(i+1)ra_{i+1}}{(n-i)!} T^i \\
&= nTA - \textstyle\sum_{i=0}^{n} \frac{a_{i-1}}{(n-i)!} T^i - \sum_{i=0}^{n-1} \frac{(n-i)(i+1)ra_{i+1}}{(n-i)!} T^i \\
&= (nT + s)A - \textstyle\sum_{i=0}^{n} \frac{1}{(n-i)!} (sa_i + a_{i-1} + (n-i)(i+1)ra_{i+1}) T^i \\
&= (nT + s)A + \textstyle\sum_{i=0}^{n} \frac{a_i'}{(n-i)!} T^i = (nT + s)A + \frac{\partial A}{\partial z}.
\end{aligned}
$$

If we now assume that the equality holds for k, i.e. that we have

$$
\frac{\partial^k A}{\partial T^k}(T^2-r) = \frac{\partial^k A}{\partial T^{k-1} \partial z} + ((n-2k+2)T+s)\frac{\partial^{k-1} A}{\partial T^{k-1}} + (k-1)(n-k+2)\frac{\partial^{k-2}A}{\partial T^{k-2}},
$$

taking in both sides the derivative with respect to T, we obtain

$$\frac{\partial^{k+1} A}{\partial T^{k+1}}(T^2 - r) + \frac{\partial^k A}{\partial T^k}(2T) = \frac{\partial^{k+1} A}{\partial T^k \partial z} + (n - 2k + 2)\frac{\partial^{k-1} A}{\partial T^{k-1}}$$

$$+((n-2k+2)T + s)\frac{\partial^k A}{\partial T^k} + (k-1)(n-k+2)\frac{\partial^{k-1} A}{\partial T^{k-1}},$$

which gives the equality for $k+1$.

To show that $y = e^{\int \omega}$ is a solution of (7.14) is equivalent to show that $\omega' + \omega^2 = r$. We assume that $\omega' + \omega^2 - r \neq 0$ and force a contradiction.

Since $A(\omega) = 0$, we have

$$\frac{\partial A}{\partial T}(\omega)\omega' + \frac{\partial A}{\partial z}(\omega) = 0.$$

Therefore

$$\frac{\partial A}{\partial T}(\omega)(\omega' + \omega^2 - r) = -\frac{\partial A}{\partial z}(\omega) + (n\omega + s)A(\omega) + \frac{\partial A}{\partial z}(\omega) = 0.$$

Hence $(\partial A/\partial T)(\omega) = 0$. We want to prove $(\partial^k A/\partial T^k)(\omega) = 0$ for all k. We have $A(\omega) = (\partial A/\partial T)(\omega) = 0$. Let us assume that

$$\frac{\partial^{k-1} A}{\partial T^{k-1}}(\omega) = \frac{\partial^k A}{\partial T^k}(\omega) = 0.$$

We then have

$$\frac{\partial^{k+1} A}{\partial T^{k+1}}(\omega)\omega' + \frac{\partial A^{k+1}}{\partial T^k \partial z}(\omega) = 0.$$

Thus

$$\frac{\partial^{k+1} A}{\partial T^{k+1}}(\omega)(\omega' + \omega^2 - r) = -\frac{\partial^{k+1} A}{\partial T^k \partial z}(\omega) + \frac{\partial^{k+1} A}{\partial T^k \partial z}(\omega)$$

$$+((n-2k)\omega + s)\frac{\partial^k A}{\partial T^k}(\omega) + k(n-k+1)\frac{\partial^{k-1} A}{\partial T^{k-1}}(\omega) = 0,$$

so $(\partial^{k+1} A/\partial T^{k+1})(\omega) = 0$. But

$$\frac{\partial^n A}{\partial T^n}(\omega) = -n! \neq 0,$$

so we have reached a contradiction which proves the proposition. □

The next two propositions give information on the degree n of the minimal polynomial over $\mathbb{C}(z)$ of an element $\omega = y'/y$ for y a solution of the differential equation.

Proposition 7.3.10. *Let y be a solution of (7.14) and $\omega = y'/y$. Let G denote the differential Galois group of the equation.*

a) If G is the tetrahedral group, then $\deg_{\mathbb{C}(z)} \omega \geq 4$ and we have equality for some solution y.

b) If G is the octahedral group, then $\deg_{\mathbb{C}(z)} \omega \geq 6$ and we have equality for some solution y.

c) If G is the icosahedral group, then $\deg_{\mathbb{C}(z)} \omega \geq 12$ and we have equality for some solution y.

Proof. With the notations in Proposition 4.6.10, we observe that the matrix $-D$ is an element of order 6 in $2A_4$, E is an element of order 8 in $2S_4$, and DBF is an element of order 10 in $2A_5$. Let y be an eigenvector of each of these matrices. Then ω is fixed by a cyclic subgroup G_1 of G of order $6, 8$, or 10 respectively, hence has degree $\leq 4, 6$, or 12 over $\mathbb{C}(z)$.

Now, for any solution y of the differential equation, let G_1 be the subgroup of G fixing ω. Completing y to a basis y, y_2 of the space of solutions and substituting G by a conjugate so that G is the Galois group of the differential equation relative to the basis y, y_2, we have that all matrices in G_1 have the form $\begin{pmatrix} c & d \\ 0 & c^{-1} \end{pmatrix}$. If G_1 has order m, then c is an mth root of unity. We want to see that G_1 is cyclic. If not, take A a matrix in G_1 of maximal order n and conjugate G_1 to make A diagonal. If B is another element in G_1, its order must be a divisor of n and, if B were not diagonal, G_1 would contain a matrix of the form $\begin{pmatrix} 1 & d \\ 0 & 1 \end{pmatrix}$ contradicting its finiteness. So G_1 is cyclic and $G_1/\{\pm 1\}$ is a cyclic subgroup of A_4, S_4, or A_5; hence its order is $\leq 3, 4$, or 5. So $|G_1| \leq 6, 8, 10$ and $\deg_{\mathbb{C}(z)} \omega \geq 4, 6, 12$. □

Proposition 7.3.11. *a) Suppose that (7.29) has a solution $s \in \mathbb{C}(z)$ for $n = 4$. Then the polynomial*

$$T^4 - \sum_{i=0}^{3} \frac{a_i}{(4-i)!} T^i \in \mathbb{C}(z)[T]$$

is irreducible over $\mathbb{C}(z)$.

b) Suppose that (7.29) has a solution $s \in \mathbb{C}(z)$ for $n = 6$. Then the polynomial

$$T^6 - \sum_{i=0}^{5} \frac{a_i}{(6-i)!} T^i \in \mathbb{C}(z)[T]$$

is irreducible over $\mathbb{C}(z)$.

c) *Suppose that (7.29) has a solution* $s \in \mathbb{C}(z)$ *for* $n = 12$ *and has no solution in* $\mathbb{C}(z)$ *for* $n = 4$ *and* $n = 6$. *Then the polynomial*

$$T^{12} - \sum_{i=0}^{11} \frac{a_i}{(12-i)!} T^i \in \mathbb{C}(z)[T]$$

is irreducible over $\mathbb{C}(z)$.

Proof. By Proposition 7.3.9, any root ω of one of the three polynomials in the statement satisfies that $e^{\int \omega}$ is a solution to $Y'' = rY$. By Proposition 7.3.10, this implies $\deg_{\mathbb{C}(z)} \omega \geq 4$. We then obtain the irreducibility of the polynomial in a) and b). Now, for c), if the polynomial had some factor of degree < 12, then, by Proposition 7.3.10 again, the differential Galois group of the equation $Y'' = rY$ would either be the tetrahedral or the octahedral group. But then, by Proposition 7.3.8, (7.29) would have a solution for $n = 4$ or $n = 6$. □

Our aim is now to find a solution to (7.29) in an effective way.

Proposition 7.3.12. *Let* F *be any homogeneous polynomial of degree* n *in solutions of (7.14). Then* $s = F'/F$ *is a solution of (7.29) for* n.

Proof. First we prove that if F_1, F_2 are elements in a differential extension of $\mathbb{C}(z)$ such that $s_1 := F_1'/F_1$ and $s_2 := F_2'/F_2$ are solutions of (7.29) for n, then $s_3 := (c_1F_1 + c_2F_2)'/(c_1F_1 + c_2F_2)$ is a solution of (7.29) for n for any $c_1, c_2 \in \mathbb{C}$. Let $a_i^1, a_i^2, a_i^3, i = n, n-1, \ldots$ denote the sequences determined by (7.29) for s_1, s_2, s_3 respectively. We claim that

$$(c_1F_1 + c_2F_2)a_i^3 = c_1F_1a_i^1 + c_2F_2a_i^2.$$

This is clear for $i = n$, since $a_n^j = -1, j = 1, 2, 3$. By induction

$$\begin{aligned}
&(c_1F_1 + c_2F_2)a_{i-1}^3 = (c_1F_1 + c_2F_2)[-a_i^{3\prime} + s_3a_i^3 - (n-i)(i+1)ra_{i+1}^3] \\
&= -[(c_1F_1 + c_2F_2)a_i^3]' - (n-i)(i+1)r(c_1F_1 + c_2F_2)a_{i+1}^3 \\
&= -[c_1F_1a_i^1 + c_2F_2a_i^2]' - (n-i)(i+1)r(c_1F_1a_{i+1}1 + c_2F_2a_{i+1}^2) \\
&= c_1F_1a_{i-1}^1 + c_2F_2a_{i-1}^2.
\end{aligned}$$

Therefore

$$(c_1F_1 + c_2F_2)a^3_{-1} = c_1F_1a^1_{-1} + c_2F_2a^2_{-1}.$$

Hence $a^3_{-1} = 0$.

Now, to prove the proposition, we may assume that

$$F = \prod_{i=1}^{n} y_i,$$

where $y_1, y_2, \ldots, y_n$ are solutions of (7.14). Let $\omega_i = y_i'/y_i$ and denote by σ_{mk} the kth symmetric function of $y_1, \ldots, y_m$. First we claim that

$$\sigma'_{mk} = (m + 1 - k)r\sigma_{m,k-1} - \sigma_{m1}\sigma_{mk} + (k + 1)\sigma_{m,k+1}.$$

For $m = 1$, the formula is clear since $\sigma_{11} = \omega_1, \sigma_{10} = 1$ and ω_1 satisfies $\omega_1' = r - \omega_1^2$. Assuming it true for $m - 1$,

$$\begin{aligned}
\sigma'_{mk} &= (\sigma_{m-1,k} + \sigma_{m-1,k-1}\omega_m)' \\
&= (m - k)r\sigma_{m-1,k-1} - \sigma_{m-1,1}\sigma_{m-1,k} + (k + 1)\sigma_{m-1,k+1} \\
&\quad +[(m + 1 - k)r\sigma_{m-1,k-2} - \sigma_{m-1,1}\sigma_{m-1,k-1} + k\sigma_{m-1,k}]\omega_m \\
&\quad +\sigma_{m-1,k-1}(r - \omega_m^2) \\
&= (m + 1 - k)r(\sigma_{m-1,k-1} + \sigma_{m-1,k-2}\omega_m) \\
&\quad -(\sigma_{m-1,1} + \omega_m)(\sigma_{m-1,k} + \sigma_{m-1,k-1}\omega_m) \\
&\quad +(k + 1)(\sigma_{m-1,k+1} + \sigma_{m-1,k}\omega_m) \\
&= (m + 1 - k)r\sigma_{m,k-1} - \sigma_{m1}\sigma_{mk} + (k + 1)\sigma_{m,k+1},
\end{aligned}$$

which completes the induction.

Next we use induction on i to prove that

$$a_i = (-1)^{n-i+1}(n - i)!\sigma_{n,n-i}.$$

For $i = n - 1$, we have $a_{n-1} = s = F'/F = \sum_{i=1}^{n} \omega_i = \sigma_{n1}$. Using (7.29), we have

$$\begin{aligned}
a_{i-1} &= -a_i' - sa_i - (n - i)(i + 1)ra_{i+1} \\
&= (-1)^{n-i}(n - i)!\sigma'_{n,n-i} + \sigma_{n1}(-1)^{n-i}(n - i)!\sigma_{n,n-i} \\
&\quad -(n - i)(i + 1)r(-1)^{n-i}(n - 1 - i)!\sigma_{n,n-1-i} \\
&= (-1)^{n-i}(n - i)![\sigma'_{n,n-i} + \sigma_{n1}\sigma_{n,n-i} - (i + 1)r\sigma_{n,n-1-i}] \\
&= (-1)^{n-i}(n - i)!(n - i + 1)\sigma_{n,n-i+1} \\
&= (-1)^{n-i}(n - i + 1)!\sigma_{n,n-i+1}.
\end{aligned}$$

Hence

$$a_{-1} = (-1)^n (n+1)! \sigma_{n,n+1} = 0.$$

□

The next proposition gives invariants of the groups $2A_4, 2S_4$, and $2A_5$.

Proposition 7.3.13. *Let G be the Galois group of (7.14) and let y_1, y_2 be a basis of the solution space relative to the group G as given in Proposition 4.6.10.*

a) If G is the tetrahedral group, then $(y_1^4 + 2\sqrt{-3}y_1^2y_2^2 + y_2^4)^3 \in \mathbb{C}(z)$.

b) If G is the octahedral group, then $(y_1^5 y_2 - y_1 y_2^5)^2 \in \mathbb{C}(z)$.

c) If G is the icosahedral group, then

$$\alpha(y_1^4 + 2\sqrt{-3}y_1^2y_2^2 + y_2^4)^3 + \beta(y_1y_2(y_1^4 - y_2^4))^2 \in \mathbb{C}(z),$$

where

$$\alpha = -5(1+i) + \sqrt{5}(-1+i), \beta = (30\sqrt{3} + 22\sqrt{5})(1+i) + (22 - 6\sqrt{15})(1-i).$$

Proof. With the notations in 4.6.10, one can check that $y_1^4 + 2\sqrt{-3}y_1^2y_2^2 + y_2^4$ is invariant by the action of the matrix B and multiplied by a cubic root of unity by the action of the matrix D. Hence $(y_1^4 + 2\sqrt{-3}y_1^2y_2^2 + y_2^4)^3$ is fixed by $2A_4$, so belongs to $\mathbb{C}(z)$. Analogously, $y_1^5y_2 - y_1y_2^5$ is invariant by the action of D and changes sign under the action of E, so $(y_1^5y_2 - y_1y_2^5)^2$ is fixed by $2S_4$, so belongs to $\mathbb{C}(z)$. Finally $y_1y_2(y_1^4 - y_2^4)$ is invariant under the action of both B and D, so $\alpha(y_1^4 + 2\sqrt{-3}y_1^2y_2^2 + y_2^4)^3 + \beta(y_1y_2(y_1^4 - y_2^4))^2$ is also for all $\alpha, \beta \in \mathbb{C}$. Now it can be checked that $\alpha(y_1^4 + 2\sqrt{-3}y_1^2y_2^2 + y_2^4)^3 + \beta(y_1y_2(y_1^4 - y_2^4))^2$ is also invariant by the action of F for the values of α and β in the statement, hence fixed by $2A_5$, so belongs to $\mathbb{C}(z)$. □

From these invariants, we obtain in each case a particular solution of (7.29) suitable for explicit determination.

Proposition 7.3.14. *a) If G is the tetrahedral group, then (7.29) has a solution $s = u'/u$, where $u^3 \in \mathbb{C}(z)$, for $n = 4$.*

b) If G is the octahedral group, then (7.29) has a solution $s = u'/u$, where $u^2 \in \mathbb{C}(z)$, for $n = 6$.

c) If G is either the tetrahedral group, the octahedral group or the icosahedral group, then (7.29) has a solution $s = u'/u$, where $u \in \mathbb{C}(z)$, for $n = 12$.

Proof. This follows from Propositions 7.3.12 and 7.3.13. □

We now write

$$u^{12/n} = \prod_{c \in \mathbb{C}} (z-c)^{e_c} \in \mathbb{C}(z),$$

where $n = 4, 6$, or 12 and $e_c \in \mathbb{Z}$. As in the other cases, the algorithm determines the possible values for e_c using local conditions, then decides which families give a global solution. We describe the algorithm in the following proposition.

Proposition 7.3.15. *Let $r \in \mathbb{C}(z)$ satisfy the necessary conditions for case 3 given in Theorem 7.3.2. Let Γ denote the set of poles of r in the complex plane. Let n be the degree of the polynomial equation for ω we are looking for.*

Step 1. For each $c \in \Gamma \cup \{\infty\}$, we define a set E_c as described below.

(c1) If $c \in \Gamma$ and c is a pole of order 1, then

$$E_c = \{12\}.$$

(c2) If $c \in \Gamma$ and c is a pole of order 2, then

$$E_c = \{6 + \frac{12k}{n}\sqrt{1+4b} \; : \; k = 0, \pm 1, \pm 2, \ldots, \pm \frac{n}{2}\} \cap \mathbb{Z},$$

for b the coefficient of $(z-c)^{-2}$ in the partial fraction expansion for r.

(∞1) If the order of r at ∞ is 2, then

$$E_\infty = \{6 + \frac{12k}{n}\sqrt{1+4b} \; : \; k = 0, \pm 1, \pm 2, \ldots, \pm \frac{n}{2}\} \cap \mathbb{Z},$$

where b is the coefficient of $1/z^2$ in the Laurent series expansion of r at ∞.

(∞2) If the order of r at ∞ is > 2, then

$$E_\infty = \{6 + \frac{12k}{n} \; : \; k = 0, \pm 1, \pm 2, \ldots, \pm \frac{n}{2}\} \cap \mathbb{Z}.$$

Step 2. We consider the families $(e_c)_{c \in \Gamma \cup \infty}$ with $e_c \in E_c$. For each such family, let

$$d = \frac{n}{12}(e_\infty - \sum_{c \in \Gamma} e_c).$$

If d is a nonnegative integer, the family is retained; otherwise it is discarded. If no families are retained, then ω cannot satisfy a polynomial equation of degree n with coefficients in $\mathbb{C}(z)$.

Step 3. For each family retained from step 2, let

$$\theta = \frac{n}{12}\sum_{c\in\Gamma}\frac{e_c}{z-c}, \quad S = \prod_{c\in\Gamma}(z-c).$$

Next search for a polynomial $P \in \mathbb{C}(z)$ of degree d such that when we define polynomials $P_n, P_{n-1}, \dots, P_{-1}$ recursively by the formulas below, then P_{-1} is identically zero.

$$\begin{aligned} &P_n = -P \\ (7.30)\quad &P_{i-1} = -SP_i' + ((n-i)S' - S\theta)P_i - (n-i)(i+1)S^2rP_{i+1}, \\ &\qquad\qquad (i = n, n-1, \dots, 0). \end{aligned}$$

If such a polynomial exists for some (e_c), let ω be a root of the polynomial

$$\sum_{i=0}^{n}\frac{S^iP_i}{(n-i)!}T^i = 0.$$

Then $y = e^{\int\omega}$ is a solution to (7.14).

If no polynomial P is found for any family retained from step 2, then (7.14) has no solution of the form $y = e^{\int\omega}$, with ω algebraic of degree n over $\mathbb{C}(z)$.

Proof. We first determine the sets E_c of possible values of e_c in step 1. For ease of notation we assume $c = 0$ and write $e = e_0$. We use the Laurent expansions for

$$s = \frac{u'}{u} = \frac{n}{12}\frac{(u^{12/n})'}{u^{12/n}} = \frac{n}{12}ez^{-1} + \dots$$

and for r, namely

$$r = b_{-2}z^{-2} + b_{-1}z^{-1} + \dots,$$

with $b_{-2}, b_{-1} \in \mathbb{C}$. Note that by the necessary conditions in Proposition 7.3.2, r has no pole of order > 2. The proof is more involved than in the other cases. We split it in several lemmas.

Lemma 7.3.16. *If $b_{-2} = 0$ and $b_{-1} \neq 0$, then $e = 12$.*

Proof. We write

$$s = \frac{n}{12}ez^{-1} + f + \dots$$

and treat e and f as indeterminates. Then, using (7.29), we can write

$$a_i = A_i z^{i-n} + B_i z^{i-n+1} + C_i f z^{i-n+1} + \dots, \tag{7.31}$$

where A_i, B_i, C_i are polynomials in e with coefficients in $\mathbb{C}$ satisfying the following recursive relations.

$$\begin{aligned}
A_n &= -1, \quad A_{i-1} = (n - i - \frac{n}{12}\,e)\,A_i,\\
B_n &= 0, \qquad B_{i-1} = (n - i - 1 - \frac{n}{12}\,e)\,B_i - (n-i)(i+1)b_{-1}A_{i+1},\\
C_n &= 0, \qquad C_{i-1} = (n - i - 1 - \frac{n}{12}\,e)\,C_i - A_i,
\end{aligned}$$

for $i = n, \dots, 0$. The solution to these equations is given by

$$\begin{aligned}
A_i &= -\prod_{j=0}^{n-i-1}\left(j - \frac{n}{12}e\right)\\
B_i &= b_{-1}\sum_{j=0}^{n-i-2}(j+1)(n-j)\prod_{\substack{k=0\\k\neq j}}^{n-i-2}\left(k - \frac{n}{12}e\right)\\
C_i &= (n-i)\prod_{j=0}^{n-i-2}\left(j - \frac{n}{12}e\right), i = n, \dots, 0.
\end{aligned}$$

Since $0 = a_{-1} = A_{-1}z^{-n-1} + B_{-1}z^{-n} - C_{-1}fz^{-n} + \dots$, we obtain

$$0 = A_{-1} = -\prod_{j=0}^{n}(j - \frac{n}{12}e)$$

and

$$\begin{aligned}
0 &= B_{-1} + C_{-1}f\\
&= b_{-1}\sum_{j=0}^{n-1}(j+1)(n-j)\prod_{\substack{k=0\\k\neq j}}^{n-1}\left(k - \frac{n}{12}e\right) + f(n+1)\prod_{k=0}^{n-1}\left(k - \frac{n}{12}e\right).
\end{aligned}$$

The first equation implies that

$$e = \frac{12}{n} l,$$

for some $l = 0, \ldots, n$. Suppose that $l \neq n$. Then the second equation gives

$$0 = b_{-1}(l+1)(n-l) \prod_{\substack{k=0 \\ k \neq l}}^{n-1} (k-l),$$

which implies that $b_{-1} = 0$. This contradiction shows that $l = n$ and therefore $e = 12$. □

Lemma 7.3.17. *If $b_{-2} \neq 0$, then e is an integer chosen from among*

a) $6 + k\sqrt{1+4b_{-2}}$, $k = 0, \pm 3, \pm 6$ if $n = 4$,

b) $6 + k\sqrt{1+4b_{-2}}$, $k = 0, \pm 2, \pm 4, \pm 6$ if $n = 6$,

c) $6 + k\sqrt{1+4b_{-2}}$, $k = 0, \pm 1, \ldots, \pm 6$ if $n = 12$.

Proof. Writing again a_i as in (7.31) and using (7.29), we obtain

$$\begin{aligned} A_n &= -1, \\ A_{i-1} &= \left(n - i - \frac{n}{12} e\right) A_i - (n-i)(i+1) b_{-2} A_{i+1}. \end{aligned}$$

If y is a solution to (7.14), $y = z^{\mu} + \ldots$ its Puiseux series expansion, we have $\mu(\mu - 1) = b_{-2}$. Assuming $b_{-2} \neq 1/4$, the differential equation has Puiseux series solutions of the form

$$\begin{aligned} y_1 &= z^{\mu_1} + \ldots \quad \text{where } \mu_1 = (1 + \sqrt{1+4b_{-2}})/2, \\ y_2 &= z^{\mu_2} + \ldots \quad \text{where } \mu_2 = (1 - \sqrt{1+4b_{-2}})/2. \end{aligned}$$

By Proposition 7.3.12, $(y_1^i y_2^{n-i})'/(y_1^i y_2^{n-i})$ is a solution of (7.29) for n. Since

$$\frac{(y_1^i y_2^{n-i})'}{y_1^i y_2^{n-i}} = (i\mu_1 + (n-i)\mu_2) z^{-1} + \cdots = \left(\frac{n}{2} - \left(\frac{n}{2} - i\right) \sqrt{1+4b_{-2}}\right) z^{-1} + \ldots$$

the polynomial A_{-1} must vanish for

$$\frac{n}{12} e = \frac{n}{2} - \left(\frac{n}{2} - i\right) \sqrt{1+4b_{-2}}. \tag{7.32}$$

a) For $n = 4$ and $b_{-2} \neq 1/4$, we obtain from (7.32), $e = 6 + k\sqrt{1+4b_{-2}}$, $k = 0, \pm 3, \pm 6$. If $b_{-2} = -1/4$, by direct computation we obtain $A_{-1} = (e-6)^5/243$.

b) For $n = 6$ and $b_{-2} \neq 1/4$, we obtain from (7.32), $e = 6 + k\sqrt{1 + 4b_{-2}}$, $k = 0, \pm 2, \pm 4, \pm 6$. If $b_{-2} = -1/4$, by direct computation we obtain $A_{-1} = (e-6)^7/128$.

c) For $n = 12$ and $b_{-2} \neq 1/4$, we obtain from (7.32), $e = 6 + k\sqrt{1 + 4b_{-2}}$, $k = 0, \pm 1, \ldots, \pm 6$. If $b_{-2} = -1/4$, by direct computation we obtain $A_{-1} = (e-6)^{13}$.

□

Lemma 7.3.18. *If $b_{-2} = b_{-1} = 0$, i.e. at an ordinary point of r, we have that $(n/12)e$ is an integer.*

Proof. As in the proof of Lemma 7.3.16, we obtain

$$e = \frac{12}{n} l,$$

for some $l = 0, \ldots, n$. □

Let Γ denote the set of poles of r. For $c \in \Gamma$, let b_c denote the coefficient of $(z-c)^{-2}$ in the partial fraction expansion for r. Until now, we have proved the following.

(1) In the tetrahedral case, (7.29) has a solution $s = u'/u$, for $n = 4$, where

$$u^3 = P^3 \prod_{c \in \Gamma} (z-c)^{e_c},$$

$P \in \mathbb{C}[z]$ and $e_c \in \{6 + k\sqrt{1 + 4b_c} : k = 0, \pm 3, \pm 6\} \cap \mathbb{Z}$.

(2) In the octahedral case, (7.29) has a solution $s = u'/u$, for $n = 6$, where

$$u^2 = P^2 \prod_{c \in \Gamma} (z-c)^{e_c},$$

$P \in \mathbb{C}[z]$ and $e_c \in \{6 + k\sqrt{1 + 4b_c} : k = 0, \pm 2, \pm 4, \pm 6\} \cap \mathbb{Z}$.

(3) In either the tetrahedral case, the octahedral case, or the icosahedral case, (7.29) has a solution $s = u'/u$, for $n = 12$, where

$$u = P \prod_{c \in \Gamma} (z-c)^{e_c},$$

$P \in \mathbb{C}[z]$ and $e_c \in \{6 + k\sqrt{1 + 4b_c} : k = 0, \pm 1, \ldots, \pm 6\} \cap \mathbb{Z}$.

Let d be the degree of the polynomial P. Then the Laurent series for s at ∞ has the form

$$s = \frac{n}{12}\left(\frac{12}{n}d + \sum_{c\in\Gamma} e_c\right) z^{-1} + \dots$$

and the Laurent series for r at ∞ has the form

$$r = \gamma z^{-2} + \dots$$

since by the necessary conditions in Proposition 7.3.2, the order of r at ∞ is at least 2.

If we let

$$e_\infty = \frac{12}{n}d + \sum_{c\in\Gamma} e_c,$$

then it can be proved as in Lemma 7.3.17 that e_∞ satisfy the conditions in the statement of Proposition 7.3.15. Also

$$d = \frac{n}{12}\left(e_\infty - \sum_{c\in\Gamma} e_c\right)$$

must be a nonnegative integer. This is the justification of step 2 of the algorithm.

We shall complete the proof of the algorithm by showing that the recursive relations of step 3 are identical with (7.29). Let

$$\theta = \frac{n}{12}\sum_{c\in\Gamma}\frac{e_c}{z-c} \quad \text{and} \quad S = \prod_{s\in\Gamma}(z-c).$$

Then $s = u'/u = P'/P + \theta$. Also set $P_i = S^{n-i}Pa_i$. Using (7.29), we have

$$\begin{aligned}
P_n &= -P \\
P_{i-1} &= S^{n-i+1}Pa_{i-1} \\
&= S^{n-i+1}P(-a_i' - sa_i - (n-i)(i+1)ra_{i+1}) \\
&= -S(S^{n-i}Pa_i)' + (n-i)S^{n-i}S'Pa_i + S^{n-i+1}P'a_i \\
&\quad -S(P'+P\theta)(S^{n-i}a_i) - (n-i)(i+1)S^2r(S^{n-i-1}Pa_{i+1}) \\
&= -SP_i' + ((n-i) - S\theta)P_i - (n-i)(i+1)S^2rP_{i+1}.
\end{aligned}$$

This is exactly (7.30) in step 3 of the algorithm. Finally the equation

$$\omega^n = \sum_{i=0}^{n-1} \frac{a_i}{(n-i)!}\omega^i$$

may be rewritten as

$$0 = -S^n P \omega^n + \sum_{i=0}^{n-1} \frac{S^n P a_i}{(n-i)!}\omega^i = \sum_{i=0}^{n} \frac{S^i P_i}{(n-i)!}\omega^i.$$

□

Example 7.3.19. The computations in these examples have been made with Maple.

1. Let us again consider the equation in Example 7.3.3.**3**. The poles of r are $c_1 = i/\sqrt{3}, c_2 = -i/\sqrt{3}$ of order 2. Following the notations in Proposition 7.3.15, with $n = 4$, we obtain $E_{c_1} = E_{c_2} = \{4, 5, 6, 7, 8\}, E_\infty = \{3, 6, 9\}$. The only nonnegative integer value of $d = (e_\infty - e_{c_1} - e_{c_2})/3$ is 0, with $e_\infty = 9, \{e_{c_1}, e_{c_2}\} = \{4, 5\}$. We choose $e_{c_1} = 5, e_{c_2} = 4$ and put

$$\theta = \frac{9z + i/\sqrt{3}}{3z^2+1}, \qquad S = z^2 + 1/3.$$

We obtain that the differential equation has a solution of the form $\eta = e^{\int \omega}$, for ω a root of the irreducible polynomial

$$\sum_{i=0}^{4} \frac{S^i P_i}{(4-i)!} T^i,$$

where

$$\begin{aligned}
P_4 &= -1 \\
P_3 &= 3z + \frac{\sqrt{3}i}{9} \\
P_2 &= \frac{27}{4}z^2 - \frac{4\sqrt{3}i}{9}z - \frac{1}{36} \\
P_1 &= \frac{81}{8}z^3 + \frac{7\sqrt{3}}{8}z^2 + \frac{25}{216}z - \frac{\sqrt{3}i}{216} \\
P_0 &= -\frac{243}{32}z^4 - \frac{3\sqrt{3}i}{4}z^3 - \frac{71}{432}z^2 + \frac{13\sqrt{3}i}{972}z - \frac{1}{864}.
\end{aligned}$$

The second choice of e_{c_1}, e_{c_2} gives the conjugate of this polynomial by $i\sqrt{3} \mapsto -i\sqrt{3}$.

2. Let us again consider the equation in Example 7.3.3.**4**. The poles of r are $0, 1$ of order 2. Following the notations in Proposition 7.3.15, with $n = 4$, we obtain $E_0 = \{6\}, E_1 = \{3, 6, 9\}$ and $E_\infty = \{4, 5, 6, 7, 8\}$, so $d = (e_\infty - e_{c_1} - e_{c_2})/3$ is always negative. For $n = 6$, we obtain $E_0 = \{5,6,7\}, E_1 = \{3,4,5,6,7,8,9\}$ and $E_\infty = \{4,6,8\}$, so the only nonnegative integer value of $d = (e_\infty - e_{c_1} - e_{c_2})/2$ is 0 which occurs for $e_0 = 5, e_1 = 3, e_\infty = 8$. We then put

$$\theta = \frac{8z-5}{2z(z-1)}, \qquad S = z(z-1)$$

and obtain that the differential equation has a solution of the form $\eta = e^{\int \omega}$, for ω a root of the irreducible polynomial

$$\sum_{i=0}^{6} \frac{S^i P_i}{(6-i)!} T^i,$$

where

$$P_6 = -1$$

$$P_5 = 4z - \frac{5}{2}$$

$$P_4 = -\frac{40}{3}z^2 + \frac{1595}{96}z - \frac{165}{32}$$

$$P_3 = \frac{320}{9}z^3 - \frac{19085}{288}z^2 + \frac{11815}{288}z - \frac{135}{16}$$

$$P_2 = -\frac{640}{9}z^4 + \frac{705}{4}z^3 - \frac{250795}{1536}z^2 + \frac{154405}{2304}z - \frac{5265}{512}$$

$$P_1 = \frac{2560}{27}z^5 - \frac{5275}{18}z^4 + \frac{1664435}{4608}z^3 - \frac{6136255}{27648}z^2 + \frac{4895}{72}z - \frac{8505}{1024}$$

$$P_0 = -\frac{5120}{81}z^6 + \frac{18950}{81}z^5 - \frac{4973285}{13824}z^4 + \frac{390462455}{1327104}z^3 - \frac{179120705}{1327104}z^2$$
$$+\frac{539485}{16384}z - \frac{54675}{16384}.$$

3. Let us again consider the equation in Example 7.3.3.**5**. The poles of r are $c_1 = 1/\sqrt{5}, c_2 = -1/\sqrt{5}$ of order 2. With $n = 4$, we obtain $E_{c_1} = E_{c_2} = \{6\}, E_\infty = \{3,6,9\}$, so $d = (e_\infty - e_{c_1} - e_{c_2})/3$ is always negative. With $n = 6$, we obtain $E_{c_1} = E_{c_2} = \{6\}, E_\infty = \{3,4,5,6,7,8,9\}$, so again $d = (e_\infty - e_{c_1} - e_{c_2})/2$ is always negative. With $n = 12$, we obtain $E_{c_1} = \{4,6,8\}, E_{c_2} = \{5,6,7\}, E_\infty = \{3,4,5,6,7,8,9\}$, so the only nonnegative value of $d = e_\infty - e_{c_1} - e_{c_2}$ is 0, which occurs for $e_{c_1} = 4, e_{c_2} = 5, e_\infty = 9$. We put

$$\theta = \frac{9z - 1/\sqrt{5}}{z^2 - 1/5}, \quad S = z^2 - 1/5.$$

We obtain that the differential equation has a solution of the form $\eta = e^{\int \omega}$, for ω a root of the irreducible polynomial

$$\sum_{i=0}^{12} \frac{S^i P_i}{(12-i)!} T^i,$$

where

$$\begin{aligned}
P_{12} &= -1 \\
5P_{11} &= 45z - \sqrt{5} \\
5 \cdot 10^2 P_{10} &= 11(-3375z^2 + 152\sqrt{5}z - 5) \\
5 \cdot 10^3 P_9 &= 3^2 \cdot 11(28125z^3 - 1925\sqrt{5}z^2 + 135z + \sqrt{5}) \\
2 \cdot 10^4 P_8 &= 3^2 \cdot 11(-759375z^4 + 70200\sqrt{5}z^3 - 7826z^2 - 88\sqrt{5}z + 17) \\
5^2 \cdot 10^4 P_7 &= 3^2 \cdot 11(56953125z^5 - 6665625\sqrt{5}z^4 + 1044750z^3 \\
&\quad +12794\sqrt{5}z^2 - 5923z + 79\sqrt{5}) \\
5 \cdot 10^6 P_6 &= 3^2 \cdot 11(-5980078125z^6 + 850500000\sqrt{5}z^5 - 174943125z^4 \\
&\quad -1917760\sqrt{5}z^3 + 1725505z^2 - 47200\sqrt{5}z + 2113) \\
5^2 \cdot 10^7 P_5 &= 3^3 \cdot 7 \cdot 11(64072265625z^7 - 10764140625\sqrt{5}z^6 \\
&\quad +2778890625z^5 + 21714375\sqrt{5}z^4 - 39685405z^3 \\
&\quad +1675317\sqrt{5}z^2 - 151485z + 1045\sqrt{5}) \\
5 \cdot 10^9 P_4 &= 3^3 \cdot 7 \cdot 11(-4805419921875z^8 + 934031250000\sqrt{5}z^7 \\
&\quad -293245312500z^6 - 917550000\sqrt{5}z^5 + 5443088750z^4 \\
&\quad -316466320\sqrt{5}z^3 + 43387980z^2 - 601616\sqrt{5}z + 17165) \\
5^2 \cdot 10^9 P_3 &= 3^4 \cdot 7 \cdot 11(24027099609375z^9 - 5317998046875\sqrt{5}z^8 \\
&\quad +1983107812500z^7 - 5621062500\sqrt{5}z^6 - 44473713750z^5 \\
&\quad +3354864430\sqrt{5}z^4 - 620549260z^3 + 12975420\sqrt{5}z^2 \\
&\quad -742865z + 3733\sqrt{5}) \\
5 \cdot 10^{11} P_2 &= 3^4 \cdot 7 \cdot 11(-1081219482421875z^{10} + 269103515625000\sqrt{5}z^9 \\
&\quad -117017314453125z^8 + 1154857500000\sqrt{5}z^7 \\
&\quad +3002510531250z^6 - 283800362000\sqrt{5}z^5 + 66473149190z^4 \\
&\quad -1863698592\sqrt{5}z^3 + 160581937z^2 \\
&\quad -1619672\sqrt{5}z + 38415)
\end{aligned}$$

$$\begin{aligned}
5\cdot 10^{12}P_1 &= 3^4\cdot 7\cdot 11(16218292236328125z^{11}-4493067626953125\sqrt{5}z^{10}\\
&\quad +2244771826171875z^9-399728355468755\sqrt{5}z^8\\
&\quad -63055095468750z^7+7315325778750\sqrt{5}z^6\\
&\quad -2085651170250z^5+73532772890\sqrt{5}z^4\\
&\quad -8475909695z^3+128675127\sqrt{5}z^2-6136777z+29201\sqrt{5})\\
5\cdot 10^{14}P_0 &= 3^5\cdot 7\cdot 11(-4054573059082031 25z^{12}\\
&\quad +1239798339843750 00\sqrt{5}z^{11}-70303293457031250z^{10}\\
&\quad +1863791015625000\sqrt{5}z^9+2080083701953125z^8\\
&\quad -292962854250000\sqrt{5}z^7+98995669922500z^6\\
&\quad -4215079100720\sqrt{5}z^5+609351970405z^4-12374529096\sqrt{5}z^3\\
&\quad +889866670z^2-8527480\sqrt{5}z+205195).
\end{aligned}$$

Exercises

(1) Consider the second order differential equation

$$Y'' + a_1(z)\,Y' + a_2(z)Y = 0. \tag{7.33}$$

a) Compute the coefficients of the differential equation obtained from (7.33) by making the change of variables $z = 1/x$.

b) Express the conditions for $z = \infty$ to be a regular point for (7.33) in terms of a_1, a_2.

(2) Consider the differential equation

$$\frac{d^nY}{dz^n} + a_1(z)\,\frac{d^{n-1}Y}{dz^{n-1}} + \cdots + a_{n-1}(z)\frac{dY}{dz} + a_n(z)Y = 0 \tag{7.34}$$

and set

$$\frac{d^nY}{dx^n} + b_1(x)\,\frac{d^{n-1}Y}{dx^{n-1}} + \cdots + b_{n-1}(x)\frac{dY}{dx} + b_n(x)Y = 0 \tag{7.35}$$

the differential equation obtained from (7.34) by making the change of variables $z = 1/x$. Prove that the two following conditions are equivalent.

(a) $lim_{z\to\infty} z^i a_i$ exists and is finite for all $i = 1 \ldots n$.

(b) $lim_{x\to 0} x^i b_i$ exists and is finite for all $i = 1 \ldots n$.

(3) Consider the differential equation

$$Y'' - \frac{1+z}{z}\,Y' + \frac{1}{z}Y = 0. \tag{7.36}$$

a) Compute the local exponents and the solutions at the regular singular point $z = 0$.

b) Check that the general solution of the equation at the neighborhood of $z = 0$ is holomorphic at $z = 0$. A singular point where the general solution of the differential equation is holomorphic is called an *apparent singularity*.

(4) Prove the claim in Remark 7.1.5.
Hint: By a change of variable, one can assume that the three singular points of the equation are $0, 1, \infty$.

(5) Consider the Chebyshev differential equation

$$(1 - z^2)Y'' - zY' + \alpha^2 Y = 0,$$

where $\alpha \in \mathbb{R}$. Check that its singular points are regular. Give the local exponents at each singular point. Find the solutions as power series in the neighborhood of $z = 0$.

(6) Consider the Legendre differential equation

$$(1 - z^2)Y'' - 2zY' + n(n+1)Y = 0,$$

where $n \in \mathbb{N}$. Check that its singular points are regular. Give the local exponents at each singular point. Find the solutions as power series in the neighborhood of $z = 0$.

(7) Provide the details of the transformation of a Fuchsian differential equation of order 2 with three singular points into a hypergeometric equation outlined in Remark 7.2.4.

(8) Using Theorem 7.3.2 determine which of the cases in Theorem 7.3.1 can occur for Bessel's equation

$$Y'' = \frac{4(n^2 - z^2) - 1}{4z^2} Y, \; n \in \mathbb{C}.$$

(9) Using Theorem 7.3.2 determine which of the cases in Theorem 7.3.1 can occur for Weber's equation

$$Y'' = (\frac{1}{4}z^2 - \frac{1}{2} - n)\, Y, \; n \in \mathbb{C}.$$

(10) Using Theorem 7.3.2 determine which of the cases in Theorem 7.3.1 can occur for Legendre's equation

$$Y'' + \frac{2z}{z^2 - 1} Y' - \frac{n(n+1)}{z^2 - 1} Y, \; n \in \mathbb{N}.$$

For the remaining exercises, the reader may consult [**Kov**].

(11) Apply Kovacic's algorithm to the differential equation $Y'' = rY$, where

$$r = z^2 - 2z + 3 + \frac{1}{z} + \frac{7}{4z^2} - \frac{5}{z^3} + \frac{1}{z^4}.$$

(12) Apply Kovacic's algorithm to the Bessel's equation

$$Y'' = \left(\frac{4n^2 - 1}{4z^2} - 1\right) Y, \; n \in \mathbb{C}.$$

(13) Apply Kovacic's algorithm to the differential equation $Y'' = rY$, where r is a polynomial of degree 2 in z.

(14) Apply Kovacic's algorithm to the differential equation $Y'' = rY$, where

$$r = \frac{1}{z} - \frac{3}{16z^2}.$$

(15) Apply Kovacic's algorithm to the differential equation $Y'' = rY$, where

$$r = -\frac{3}{16z^2} - \frac{2}{9(z-1)^2} + \frac{3}{16z(z-1)}.$$

(16) Apply Kovacic's algorithm to the differential equation $Y'' = rY$, where

$$r = -\frac{5z+27}{36(z-1)^2}.$$

Chapter 8

Suggestions for Further Reading

In this last chapter, we briefly describe some of the topics in differential Galois theory and related areas in which active research is being performed.

1. In his lecture at the 1966 International Congress of Mathematicians [**Ko3**], E. Kolchin raised two important problems in Picard-Vessiot theory.
1. Given a linear differential equation $\mathcal{L}(Y) = 0$ over a differential field K, determine its Galois group (*direct problem*).
2. Given a differential field K, with field of constants C, and a linear algebraic group G defined over C, find a linear differential equation defined over K with Galois group G (*inverse problem*).

Regarding the direct problem, Kovacic's algorithm, presented in Section 7.3, determines in particular the differential Galois group of a homogeneous linear differential equation of order 2. An algorithm to determine the differential Galois group of a homogeneous linear differential equation of order 3 has been given by Singer and Ulmer in [**S-U**] and for order 4 by Hessinger [**Hes**] and later completed by Hartmann [**Ha1**].

Singer [**S**] presents a very good survey on direct and inverse problems in differential Galois theory. Later results were given by Mitschi-Singer, who solved the inverse problem for connected linear groups, and Hartmann. (See [**M-S1**], [**M-S2**], [**Ha2**].)

2. Some interesting topics in the analytic theory of differential equations are the Riemann-Hilbert problem, Stokes phenomenon, and generalizations of hypergeometric equations.

At the end of the 1850's, Riemann was the first to mention the problem of the reconstruction of a Fuchsian equation from its monodromy representation. Hilbert included it, with the following formulation, as the 21st problem in his list of Mathematical Problems given at the 1900 International Congress in Paris.

Prove that there always exists a linear differential equation of Fuchsian type with given singular points and a given monodromy group.

The interested reader can consult [**Bo**], [**Ż**], and [**P-S1**], as well as the bibliographies given there.

Consider a differential equation $\mathcal{L}(Y) = 0$ defined over the field $\mathbb{C}(\{z\})$ of convergent Laurent series in the variable z over the complex field and let $\hat{f} \in \mathbb{C}((z))$ be a solution of the equation, where $\mathbb{C}((z))$ denotes the field of formal Laurent series. The main theorem of the asymptotic theory of differential equations states that for a sector S at 0 with small enough opening, there exists a meromorphic function f on S with asymptotic expansion $\hat{f}$. The fact that uniqueness for f can be obtained only on a sector is known as Stokes phenomenon. (See e.g. [**P-S1**].)

Several authors have considered generalizations of Gauss hypergeometric function in one and several variables. All of them are included in the theory of A-hypergeometric functions due to Gel'fand, Graev, Kapranov, and Zelevinsky. (See [**G-K-Z**].) A-hypergeometric functions are solutions of certain partial differential systems. An independent theory of generalized hypergeometric functions has been developed by Dwork in [**Dw**].

3. At the end of the last century, Morales and Ramis used differential Galois theory to obtain nonintegrability criteria for Hamiltonian systems, which generalize classical results of Poincaré and Liapunov as well as more recent results of Ziglin. More precisely, they established that a necessary condition for the integrability of a Hamiltonian system is that the identity component of the differential Galois group of the variational equation along a particular solution, which is a linear differential equation, is abelian. (See [**Mo**] and [**Au**].) More recently, Morales, Ramis, and Simó [**M-R-S**] generalized this result by considering higher variational equations. Their criterion is being used by several authors to obtain nonintegrability of Hamiltonian systems coming from a variety of physical problems. An account of different concepts of integrability is given in [**Go**].

4. Some interesting contributions to the theory of differential fields have been made by model theorists. The proof of the existence of a differential closure for a differential field depends heavily on the use of methods of model theory. The first proof of the existence of an algorithm to determine the Galois group of a linear differential equation is model theoretical as well. (See [**Hr**].) Poizat [**P**] presents an interesting survey on the relationship between differential algebra and model theory. A more recent account of the relationship between differential Galois theory and model theory is given in [**Pi**].

5. A classical problem going back to Fuchs and Schwarz and also considered by Klein in [**Kl1**] is to determine when a given differential equation defined over $\mathbb{C}(z)$ has only algebraic solutions. In [**Sc**], Schwarz gives a complete answer to this question in the case of hypergeometric equations. The work of Schwarz has been generalized by several authors. Recently, Beukers [**Be**] has obtained a necessary and sufficient condition expressed in combinatorial terms for an A-hypergeometric system of differential equations to have a full set of algebraic solutions.

In the case when the differential equation $\mathcal{L}(Y) = 0$ is defined over $\mathbb{Q}(z)$, one can consider its reduction $\mathcal{L}_p(Y) = 0$ for almost all prime p, defined as the differential equation obtained by reducing modulo p the coefficients of $\mathcal{L}(Y) = 0$. In the 1960's, Grothendieck conjectured that the equation $\mathcal{L}(Y) = 0$ has a fundamental system of algebraic solutions, linearly independent over $\overline{\mathbb{Q}}$ if and only if $\mathcal{L}_p(Y) = 0$ has a fundamental system of algebraic solutions in $\mathbb{F}_p(z)$, linearly independent over $\mathbb{F}_p(z^p)$, for almost all prime p. This conjecture was later generalized by Katz (see [**Ka1**]) who proved it for rigid systems, that is, differential systems which are determined by their local data, i.e. its singular points and local exponents. Later, André proved it for some differential systems "coming from geometry" [**An**].

6. In the case of characteristic $p > 0$, one can consider iterative derivations, introduced by Hasse and Schmidt, which avoid the fact that a p-th power is always a constant. A Picard-Vessiot theory for iterative differential fields has been developed by Okugawa. (See [**O**].) An analogue of Grothendieck's conjecture using iterative derivations has been proved by Matzat [**Mat**].

7. The Picard-Vessiot theory presented here is Galois theory for linear differential equations. Kolchin, after some attempts to find a good differential analogue of the concept of normality for algebraic extensions, introduced the notion of strongly normal extension of differential fields. He developed a Galois theory of strongly normal extensions and associated to a strongly normal extension of differential fields its differential Galois group which has the structure of an algebraic group defined over the field of constants of the base field. He characterized Picard-Vessiot extensions as strongly normal extensions whose differential Galois group is a **linear**

algebraic group. He presented an example of strongly normal extension which is not Picard-Vessiot, the one obtained by adjoining a so-called Weierstrass element, and proved that its differential Galois group is the group of points of an elliptic curve over the field of constants.

More general nonlinear differential Galois theories have been proposed by Umemura [**U1**] and Malgrange [**Mal1**], [**Mal2**]. Malgrange theory has been further developed by Casale, who recovers the Morales-Ramis-Simó theory in the framework of Malgrange theory [**Ca**]. Heiderich [**He2**] has elaborated a generalization of Umemura theory which applies to differential, iterative differential, and difference fields. (See **6.** and **9.**)

8. Picard-Vessiot theory was developed by Kolchin under the assumption that the field of constants is algebraically closed. In [**U2**] Umemura pointed out that classical Galois theory cannot be seen as the Picard-Vessiot theory of algebraic extensions. Indeed, one can find examples of finite Picard-Vessiot extensions which are not normal (see Exercises 23 and 24 in chapter 5) and Galois algebraic extensions of differential fields which are not Picard-Vessiot (see Exercise 3 in chapter 6). Umemura introduced the notion of automorphic extension of differential fields which includes Picard-Vessiot extensions of differential fields and Galois algebraic extensions. He established a fundamental theorem for automorphic extensions, analogous to Proposition 6.3.1. An automorphic extension allows a finite extension of the constant fields. Umemura proves that the strongly normal extensions are precisely those automorphic extensions which do not add constants.

A Picard-Vessiot theory for differential fields with nonalgebraically closed field of constants has been considered by Dyckerhoff using Galois descent [**Dy**]. A Picard-Vessiot theory for formally real fields is being developed by Sowa [**So**].

9. A difference field is a field K with a distinguished automorphism ϕ. A classical example is a finite field with its Frobenius automorphism. One can consider linear difference equations, i.e. equations of the form $\mathcal{L}(Y) = \phi^n(Y) + a_{n-1}\phi^{n-1}(Y) + \cdots + a_0 Y = 0$, with $a_i \in K$. A Galois theory for difference equations has been developed which parallels differential Galois theory. (See [**P-S2**].) Important contributions to this topic are Di Vizio's proof of an analogue of Grothendieck's conjecture mentioned in point **5.** for difference equations [**Di**] and the development of an analogue of Malgrange theory mentioned in **7.** for difference equations accomplished by Granier [**Gr**].

Bibliography

[Al] L. V. Ahlfors, *Complex Analysis*, McGraw-Hill, 1966.

[An] Y. André, *Sur la conjecture des p-courbures de Grothendieck-Katz et un problème de Dwork*, Geometric aspects of Dwork theory. Vol. I, Walter de Gruyter GmbH & Co. KG, Berlin, 2004, pp. 55-112.

[A-M] M. F. Atiyah, I. G. MacDonald, *Introduction to commutative algebra*, Addison-Wesley, 1969.

[Au] M. Audin, *Les Systèmes Hamiltoniens et leur integrabilité*, Société Mathématique de France, 2001.
English version: *Hamiltonian Systems and Their Integrability*, American Mathematical Society, Société Mathématique de France, 2008.

[Ba] H. Bass, A. Buium, P. J. Cassidy, eds., *Selected works of Ellis Kolchin with commentary*, American Mathematical Society, 1999.

[Be] F. Beukers, *Algebraic A-hypergeometric functions*, Invent. Math. , **180** (2010), 589–610.

[Bi] A. Białynicki-Birula, *Zarys algebry*, Państwowe Wydawnictwo Naukowe, Warsaw, 1987.

[Bo] A. A. Bolibruch, *Holomorphic bundles, associated with linear differential equations and the Riemann-Hilbert problem*, The Stokes phenomenon and Hilbert's 16th problem, proceedings of the workshop held in Groningen, 31 May–3 June 1995, B. L. J. Braaksma, G. K. Immink, M. van der Put, eds., World Scientific, 1996, pp. 51–70.

[B] A. Borel, *Linear Algebraic Groups*, Graduate Texts in Mathematics 126, Springer, 1991.

[Ca] G. Casale, *Morales-Ramis Theorems via Malgrange pseudogroup*, Annales de l'institut Fourier, **59** no. 7 (2009), 2593–2610.

[C-H1] T. Crespo, Z. Hajto, *Introduction to differential Galois theory; with an appendix by Juan J. Morales-Ruiz*, Wydawnictwo PK, Cracow, Poland, 2007.

[C-H2] T. Crespo, Z. Hajto, *Differential Galois realization of double covers.* Ann. Inst. Fourier (Grenoble), **52** (2002), 1017–1025.

[Di] L. Di Vizio, *On the arithmetic theory of q-difference equations. The q-analogue of the Grothendieck-Katz's conjecture on p-curvatures*, Inventiones Mathematicae **150** (2002) 3, 517–578.

[D-L] A. Duval, M. Loday-Richaud, *Kovacic's Algorithm and Its Application to Some Families of Special Functions*, Applicable Algebra in Engineering, Communication and Computing **3** (1992), 211–246.

[Dw] B. Dwork, *Generalized Hypergeometric Functions*, Oxford Mathematical Monographs. Oxford University Press, London, 1990.

[Dy] T. Dyckerhoff, *Picard-Vessiot extensions over number fields*, Diplomarbeit, Fakultät für Mathematik und Informatik der Universität Heidelberg, 2005.

[E] R. Engelking, *General Topology*, PWN-Polish Scientific, 1977.

[F] O. Forster, *Lectures on Riemann Surfaces*, Springer, 1999.

[Fu] W. Fulton, *Algebraic curves*, Benjamin, 1969.

[G-K-Z] I. M. Gelfand, M. M. Kapranov, A. V. Zelevinsky, *Generalized Euler Integrals and A-Hypergeometric Functions*, Adv. Math. **84** (1990), 255-271.

[Go] P. Goldstein, *Kovalevska vs. Kovacic-two different notions of integrability and their connections*, Proceedings of the Differential Galois Theory workshop, T.Crespo, Z. Hajto, eds., Banach Center Publications 58, Warszawa 2002, pp. 63–73.

[Gr] A. Granier, *A Galois D-groupoid for q-difference equations*, Ann. Inst. Fourier, to appear.

[Ha1] J. Hartmann, *Invariants and differential Galois groups in degree four*, Proceedings of the Differential Galois Theory workshop, T. Crespo, Z. Hajto, eds., Banach Center Publications 58, Warszawa 2002, pp. 79-87.

[Ha2] J. Hartmann, *On the inverse problem in differential Galois theory*, J. Reine angew. Math. **586** (2005), 21–44.

[He] F. Heiderich, *Picard-Vessiot Theorie für lineare partielle Differentialgleichungen*, Diplomarbeit, Fakultät für Mathematik und Informatik der Universität Heidelberg, 2007.

[He2] F. Heiderich, *Galois theory of module fields*, PhD thesis, Universitat de Barcelona, 2010.

[Hes] S. Hessinger, *Computing the Galois group of a linear differential equation of order four*, Appl. Algebra Eng. Commun. Comput. **11** (2001), 489–536.

[Hr] E. Hrushovski, *Computing the Galois group of a linear differential equation*, Proceedings of the Differential Galois Theory workshop, T. Crespo, Z. Hajto, eds., Banach Center Publications 58, Warszawa 2002, pp. 97–138.

[Hu] K. Hulek, *Elementary algebraic geometry*, Student Mathematical Library vol. 20, American Mathematical Society, 2003.

[H] J. E. Humphreys, *Linear Algebraic Groups*, Graduate Texts in Mathematics 21, Springer, 1981.

[Hus] D. Husemöller, *Elliptic Curves*, Springer, 1987.

[I] E. L. Ince, *Ordinary differential equations*, Dover Publications, 1956.

[K] I. Kaplansky, *An introduction to differential algebra*, Hermann, 1976.

[Ka1] N. Katz, *A conjecture in the arithmetic theory of differential equations*, Bull. Soc. Math. France **110** (1982), 203–239; correction: Bull. Soc. Math. France **110** (1982), 347–348.

[Ka2] N. Katz, *Rigid local systems*, Annals of Math. Studies 139, Princeton University Press, 1996.

[Kl1] F. Klein, *Vorlesungen über das Ikosaeder und die Auflösung der Gleichungen vom fünften Grade*, Birkhäuser, 1993.
English version: *Lectures on the icosahedron and the solution of equations of the fifth degree*, Dover, 2003.

[Kl2] F. Klein, *Vorlesungen über die hypergeometrische Funktion*, reprint, Springer-Verlag, 1981.

[Kle] A. Kleshchev, *Lectures on Algebraic Groups*, available at:
http://darkwing.uoregon.edu/~klesh/teaching/AGLN.pdf.

[Ko1] E. R. Kolchin, *Algebraic matric groups and the Picard-Vessiot theory of homogeneous linear ordinary differential equations*, Ann. of Math. **49** (1948), 1–42; [Ba] pp. 87–128.

[Ko2] E. R. Kolchin, *On the Galois theory of differential fields*, Amer. J. Math. **77** (1955), 868–894; [Ba] pp. 261–287.

[Ko3] E. R. Kolchin, *Some problems in differential algebra*, Proceedings of the International Congress of Mathematicians, Moscow, 1968, pp. 269–276; [Ba] pp. 357–364.

[Ko4] E. R. Kolchin, *Differential algebra and algebraic groups*, Academic Press, 1973.

[Ko5] E. R. Kolchin, *Differential algebraic groups*, Academic Press, 1985.

[Kov] J. J. Kovacic, *An algorithm for solving second order linear homogeneous differential equations*, J. Symb. Comput. **2** (1986), 3–43.

[L] S. Lang, *Algebra*, Revised Third Edition, Springer, 2005.

[M] A. R. Magid, *Lectures on Differential Galois Theory*, University Lecture Series 7, American Mathematical Society, 1997.

[Mal1] B. Malgrange, *Le groupoïde de Galois d'un feuilletage*, Monographies de l'Enseignement Math. **38** (2001), 465–501.

[Mal2] B. Malgrange, *Pseudogroupes de Lie et théorie de Galois différentielle*, Institut des Hautes Études Scientifiques, preprint, mars 2010, IHES/M/10/11.

[Ma] H. Matsumura, *Commutative Algebra*, Benjamin, 1970.

[Mat] B. H. Matzat, *Differential equations and finite groups*, Journal of Algebra **300** (2006), 623–686.

[M-S1] C. Mitschi, M. F. Singer, *Connected Linear Groups as Differential Galois Groups*, Journal of Algebra **184** (1996), 333–361.

[M-S2] C. Mitschi, M. F. Singer, *Solvable-by-Finite Groups as Differential Galois Groups*, Ann. Fac. Sci. Toulouse Math (6) **11** (2002) no. 3, 403–423.

[Mo] J. J. Morales-Ruiz, *Differential Galois Theory and non-integrability of Hamiltonian systems*, Progress in Mathematics 179, Birkhäuser, 1999.

[M-R-S] J. J. Morales-Ruiz, J-P. Ramis, C. Simó, *Integrability of Hamiltonian systems and differential Galois groups of higher variational equations*, Ann. Sci. École Norm. Sup. (4) **40** (2007), 845–884.

[N-S-T] P. M. Neumann, G. A. Stoy, E. C. Thompson, *Groups and Geometry*, Oxford University Press, 1994.

[O] K. Okugawa, *Differential Algebra of Nonzero Characteristic*, Lectures in Mathematics 16, Kinokuniya Company Ltd., Tokyo, 1987.

[Pi] A. Pillay, *Around Differential Galois Theory*, Algebraic theory of differential equations, M. A. H. MacCallum, A. V. Mikhailov, eds., Cambridge University Press, 2009, pp. 232–240.

[P] B. Poizat, *Les corps différentiellement clos, compagnons de route de la théorie des modèles*, [Ba] pp. 555–565.

[Po] E. G. C. Poole, Introduction to the theory of linear differential equations, Clarendon Press, 1936.

[P-S1] M. van der Put, M. F. Singer, *Galois Theory of Linear Differential Equations*, Grundlehren der mathematischen Wissenschaften 328, Springer, 2003.

[P-S2] M. van der Put, M. F. Singer, *Galois Theory of Difference Equations*, Lecture Notes in Mathematics 1666, Springer-Verlag, 1997.

[R] J. F. Ritt, *Differential Algebra*, Dover, 1966.

[Sch] H. A. Schwarz, *Ueber diejenigen Fälle, in welchen die Gaussische Hypergeometrische Reihe eine algebraische Function ihres vierten Elementes darstellt*, J. reine angew. Math. **75** (1873), 292–335.

[Sc] W. R. Scott, *Group theory*, Dover, 1987.

[Se] A. Seidenberg, *Contribution to the Picard-Vessiot theory of homogeneous linear differential equations*, Amer. J. Math **78** (1956), 808–817.

[Sh] I. R. Shafarevich, *Basic Algebraic Geometry 1*, Springer, 1994.

[Si] J. H. Silverman, *The arithmetic of elliptic curves*, Springer, 1986.

[S] M. F. Singer, *Direct and inverse problems in differential Galois theory*, [Ba] pp. 527–554.

[S-U] M. F. Singer, F. Ulmer, *Galois groups of second and third order differential equations*, J. Symbolic Computation **16** (1993), 9–36.

[So] E. Sowa, *Picard-Vessiot extensions for real fields*, Proc. Amer. Math. Soc., electronically published, December 2010.

[Sp] T. A. Springer, *Linear Algebraic Groups*, Progress in Mathematics 9, Birkhäuser, 1998.

[Ta-Y] P. Tauvel, R. W. T. Yu, *Lie algebras and algebraic groups*, Springer Verlag, 2005.

[T-T] C. Tretkoff, M. Tretkoff, *Solution of the inverse problem of differential Galois theory in the classical case*, Amer. J. Math. **101** (1979), 1327–1332.

[U1] H. Umemura, *Differential Galois theory of infinite dimension*, Nagoya Math. J., **144** (1996), 59-135.

[U2] H. Umemura, *Galois theory of algebraic and differential equations*, Nagoya Math. J., **144** (1996), 1–58.

[Ż] H. Żołądek, *The Monodromy Group*, Monografie Matematyczne Instytut Matematyczny PAN 67, Birkhäuser, 2006.

Index